£18.00

THE INFLUENCE OF VEGETATION ON CLAYS

THOMAS TELFORD LTD
LONDON 1984

Published for the Institution of Civil Engineers by Thomas Telford Ltd, PO Box 101, 26–34 Old Street, London EC1P 1JH.

The papers in this volume were first published as a Symposium in Print in *Géotechnique* June 1983.

ISBN 0 7277 0221 1

Printed in Great Britain by Billings and Son Ltd, Worcester

CONTENTS

Preface

The roots of vegetation extract moisture from the soil in order to support growth and, generally, the soil moisture content is modified from that prevailing where roots are absent. In clay soils, changes in moisture content result in changes in volume and consequently any foundation located in the affected soil will experience some movement, certainly vertical and possibly also in a horizontal direction. These foundation movements can be cyclical on a seasonal basis, progressive settlement where the vegetation is establishing a permanent soil moisture deficit, or progressive heave when such vegetation is later removed. Damage to the structure can be considerable and very many cases occur. It is reported that in the USA the damage due to swelling and shrinking exceeds the combined damage due to the effects of earthquakes, tornadoes, hurricanes and floods.

The capacity of vegetation to cause swelling and shrinking in clay soils, and consequently to cause damage to structures, depends on a series of factors which include vegetation, soil and groundwater conditions, climate, and foundations and their distance from the vegetation. The interaction of these factors is complicated. It is hoped that the papers in this Symposium in Print and the discussion will lead to a better understanding of the problem, and of ways by which satisfactory foundations can be constructed.

The Symposium in Print contains seven papers which describe some of the experience gained and some of the studies currently in progress in Australia, Israel, South Africa, the UK and the USA.

The UK has a mild damp climate, in comparison with the more arid climates of some of the other countries, and therefore does not experience the substantial ground movements which can occur in these arid countries. Nevertheless, in the UK, vertical movements due to large trees are often of the order of 100 mm and can cause serious damage to structures.

Previous work has suggested that the minimum distance between a structure on traditional foundations and trees should be 1–1·5*h* where *h* is the height of the trees. This guideline is often unacceptable for environmental reasons and, in fact, there are very many cases where trees are closer than these limits and have not caused damage.

The observed moisture content patterns in clay soils due to various types of vegetation are reported in some of the papers. The paper by Biddle (UK) gives observations of the pattern of moisture content for various types of trees in various typical clay soils of south-east England. The results show a marked difference in the patterns for different types of trees, those trees having the highest water demand, e.g. poplars, affecting the soil to the greatest distance and depth. The results also indicate the considerable effect on the moisture pattern of bands of permeable material within the clay, which permits a rapid replenishment of any moisture deficiency that would otherwise have developed. The papers by Richards, Peter & Emerson (Australia) and Williams & Pidgeon (South Africa) give observations of the extent of the desiccation for trees such as eucalyptus and gum which are adapted to arid conditions.

If the vegetation is removed, after it has established a moisture deficiency in clay soil, the moisture contents of the soil will now increase to new equilibrium values, thus causing heave. This has been a major cause of damage to structures which have been built on sites shortly after the vegetation has been cleared. A number of case histories are given in the papers.

The papers tend to the consensus that the tendency of a clay soil to shrink or swell can best be identified by the soil suction. This is the suction pressure of the pore water, expressed on a logarithmic scale pF, where pF is the logarithm to the base 10 of the suction head in cm. Soil suction determines the moisture content and experimental relationships between these two components can be determined for various soil types. Instruments are described for measuring soil suction in situ. It is reported that the maximum suction that can be imposed by the roots of vegetation is 4–4·5 pF. If, owing to low moisture content, the soil suction exceeds that value, the vegetation is unable to extract any further moisture from the soil and, in the absence of any alternative supply of water, the vegetation will wilt and die. The suction for wet fully drained soil is 3–3·5 pF. The relationships between soil suction, moisture content and volume change suggest a possible analytical method for estimating heave/settlement. On the basis of experience in the UK, Driscoll suggests that desiccation begins to cause damage at suctions in excess of about 3 pF, which is equivalent to a moisture content of about 0·4 times the liquid limit.

One objective of this Symposium in Print is to stimulate discussion on guidelines for the design and location of satisfactory foundations. In the UK, there is a suggestion that a modest deepening of the foundations from a traditional depth of 1 to 1·5 m would considerably reduce the chance of damage. The papers suggest that, throughout the world, the general guideline is a minimum distance, equal to the height of the tree, but that much is dependent on local experience.

T. R. M. Wakeling

The legal aspects of the influence of vegetation on the swelling and shrinking of clays

R. WYLES*

INTRODUCTION

The purpose of this article is to summarize briefly the nature of the legal responsibilities which may fall on a professional man who ignores the technical problems associated with siting buildings on clay soil too close to trees and other vegetation.

In 1947, the Building Research Station published a digest (no. 3) which stressed the importance of taking care when siting dwellings in the vicinity of trees and shrubs. Even more importantly, in 1954 the British Standards Institution published a code of practice (no. 4) which stated that . . . 'the shrinkage of clay may be increased by the drying effect produced by fast growing water seeking trees, such as poplars and elms. The range of influence depends on the site and the number of trees as it increases during dry periods, but in general it is desirable that there should be at least 25′ between such trees and a building.'

The Building Regulations (1976), in compliance with which all buildings must be constructed, state 'the foundations shall be taken down to such a depth or be so constructed as to safeguard the building against damage by swelling, shrinkage or freezing of the sub-soil.' The test of whether this regulation has been satisfied is whether the construction complies with the 1954 Code of Practice (British Standards Institution, 1954). Thus, failure to comply with this Code may give rise to liability on the part of the builder, local authority or professional man, if damage occurs (i.e. for breach of statutory duty).

THE PROFESSIONAL MAN'S DUTY OF CARE

Prior to 1976,[1] the liability of a professional man employed to design a building was prescribed by his contract, subject to a term implied by law into that contract (unless otherwise specifically excluded) that he would use reasonable care and skill in performing his contractual duties. The standard of care against which he was judged was that exercised by a competent practitioner of the same art.[2] As this duty to exercise reasonable care and skill arose in contract only, an action against him would have been statute barred unless a writ was issued within six years from the date that the contractual breach took place.

LIABILITY IN NEGLIGENCE

In 1976, the Court of Appeal[3] held that a professional man's duty was owed not only in contract but also in tort (i.e. negligence) for breach of the same duty of care. The importance of this judgement might have been overlooked but for another Court of Appeal decision[4] in the same year, which extended the liability period for actions in negligence in respect of latent damage in buildings. It was held that a plaintiff could bring an action within six years from the date when the damage was discovered or from when that damage became reasonably discoverable, irrespective of when the negligent act was performed.

Thus, after 1976, a professional man could be sued in negligence, if damage occurred consequent upon his failure to design or supervise, many years after his contract had been performed.

A recent Court of Appeal decision[5] illustrates this principle of infinite liability. A house was built in 1955 and damage appeared in 1961 and 1966, as a result of which minor repairs were carried out by the plaintiff. It was only in 1977, when more serious damage had occurred, that its cause was established, namely the unsuitability of the foundations for the ground conditions. Proceedings were started in 1978. It was held that only in 1977 could the plaintiff reasonably have discovered the damage, and that therefore his action was not statute barred.

The House of Lords ostensibly reversed this principle of infinite liability when it decided[6] that the reasonable discoverability principle was wrong, and that the limitation period for actions in negligence in respect of latent damage is six years from the date damage occurs and not from when it becomes manifest. In practice, the date when damage occurs would normally be a date sometime between when a negligent act is performed and the date when the plaintiff becomes aware of the damage.

It is the defendant who, as a matter of law, must

* Head of Legal Services, National House-building Council.

raise a defence of limitation if he is to prove that an action is statute barred. Thus, the onus is on him to prove when the damage occurred. This could entail detailed and complicated technical evidence, the cost of which might be prohibitive and certainly would be expensive.[7]

DAMAGES RECOVERABLE FOR BREACH OF CONTRACT OR IN NEGLIGENCE

In a successful action—whether for breach of contract or in negligence—the plaintiff will normally recover damages which are intended to put him in the position he would have been in had the breach of contract or the negligent act not occurred. This means that he can normally recover the cost of repairs, together with any depreciation in value (i.e. the difference in market value between the building as repaired and the building if it had been properly constructed), plus the cost of equivalent accommodation whilst repairs are carried out. Consequential losses, including loss of business or rent, might also be recoverable. If repairs are carried out prior to judgement, the plaintiff can recover that sum plus interest until judgement. If repairs are delayed until judgement, the plaintiff will be awarded the cost of repairs as at the date of judgement. A defendant who therefore defends proceedings for a number of years will almost certainly pay more than if he conceded liability when the action was first brought.

CONTRIBUTION FROM THIRD PARTIES

While it is no defence for a professional man to argue that he relied on the recommendations or instructions of a third party, negligence by that third party may, however, reduce the amount of damages awarded against him (either because that third party is a joint tortfeasor or he is liable under the Civil Liability Contribution Act 1978).

The case of Acrecrest[8] is a good illustration of this, particularly as it also deals with the influence of vegetation on clay soils. In 1971 the plaintiff development company employed the architect defendants to design a block of flats. The architects specified 3′ 6″ foundations; the local authority building control inspector, having found tree roots on the site, specified that in that part of the site the foundations could be 5′ but elsewhere they need only be between 3′ 6″ and 4′ deep. The foundations were built strictly in accordance with these instructions but in 1973 cracks appeared as a result of heave. The architects admitted negligence and agreed the amount of damages. They then continued the proceedings against the local authority for a contribution on the ground that the local authority, by specifying the foundation depth, had assumed responsibility for the design.

The soil of the building site was clay; on the site, fruit trees had been removed by the builders while nearby poplars had been pollarded and elm trees, after construction of the flats, had been felled as a result of Dutch elm disease. Heave was caused by the clay absorbing water which had previously been absorbed by the nearby trees.

It was held that the duty of the architect (and also of the local authority) was to consider the presence and proximity of the trees in relation to the foundation design. The fact that the elm trees died later from a disease which could not have been anticipated at the time of construction did not reduce the architects' responsibility because they should have been aware of the danger of subsidence. Liability was shared in the proportion 75% (architects) : 25% (local authority).

In another similar case[9] the fact that the plaintiff himself cut down the nearby trees, after completion of the construction, thus causing heave, was held to be irrelevant.

When considering a claim for contribution from a third party, it should be noted that a professional man may be under a higher duty than, for instance, a local authority; thus, a local authority was held not to be liable when it relied on the expertise of consultant structural engineers whose recommendations had been incorporated in the foundation design.[10]

DAMAGE OCCURS—ARE REPAIRS NECESSARY?

The introduction of subsidence cover in standard form house owners insurance policies in 1971 resulted in a large number of subsidence claims during the 1970s (20 972 in 1976). As a result, the Building Research Establishment (BRE) carried out research into the necessity for extensive repairs where subsidence occurred. This resulted in the publication of BRE Digest 251 (Building Research Establishment, 1981). It concluded that many professional advisers involved in advising on repairs do not record properly the amount of damage which occurs, nor do they monitor that damage to establish whether or not it is progressive. While shrinkage and heave due to clay soils were found by BRE to be the major cause of damage to foundations, in the majority of cases such damage rarely necessitates more than cosmetic repairs, and in all cases, monitoring is imperative.

Recently a local authority was held not to be liable for the cost of underpinning which had been carried out on the advice of a structural engineer, as the damage was only slight and no monitoring had been carried out.[11]

BRE Digest 251 is important, therefore, not only for its technical conclusions, but also because it may afford a defence as to quantum in any claim relating to foundation damage in dwellings, and any professional man who ignores its conclusions may, in the future, be at risk if he specifies unnecessary and expensive repairs.

CONCLUSION

Clearly no professional man should ignore current technical literature and, in particular, any publications emanating from authoritative sources such as the Building Research Establishment, British Standards Institution or Building Research Station. Such technical literature will usually set the standard against which a professional man is likely to be judged if damage occurs For this reason, the papers in this Symposium in Print may prove invaluable in determining what a professional man should consider when employed to design buildings on clay soils which are likely to be affected by vegetation.

CASES

1. *Bagot* v. *Stevens Scanlan* (1964). 2 Building Law Reports.
2. *Bolam* v. *Friern Hospital Management Committee* (1957). 1 Weekly Law Reports.
3. *Esso Petroleum Co. Ltd* v. *Mardon* (1976). 2 Building Law Reports.
4. *Sparham-Souter and Ryan* v. *Town and Country Developments (Essex) Ltd* (1976). 3 Building Law Reports.
5. *Dennis* v. *Charnwood Borough Council* (1982). 3 All England Law Reports.
6. *Pirelli General Cable Works Limited* v. *Oscar Faber & Partners* (1983). 1 All England Law Reports.
7. Note: The Law Reform Committee Sub-Committee on Latent Damage is expected to report soon. Following Pirelli, its recommendations are likely to be a limitation period based on reasonable discoverability as in Sparham-Souter, with a longstop on liability of between 10 and 20 years.
8. *Acrecrest Ltd* v. *W. S. Hattrell & Partners* (1983). 1 All England Law Reports.
9. *Oxborrow* v. *Godfrey Davis (Artscape) Ltd* (1980).
10. *Stewart* v. *Cambridgeshire County Council* (1982).
11. *Bluett and Another* v. *Woodspring District Council* (1982). Estates Times, May.

REFERENCES

British Standards Institution (1954). Code of practice. No. 4. London: British Standards Institution.
Building Regulations (1976). London: HMSO.
Building Research Establishment (1981). *Assessment of damage in low-rise buildings, with particular reference to progressive foundation movement*. Digest 251. London: HMSO.
Building Research Station (1947). *Foundations on shrinkable clays*. Digest 3. London: HMSO.

The influence of vegetation on the swelling and shrinking of clay soils in Britain

R. DRISCOLL*

The Paper describes volume change properties of the more important British clay soils. Seasonal variations in soil moisture and the significant reductions that trees can induce are discussed. The impact of the 1975–76 drought in Britain which generated many cases of building damage is considered by the application of a damage classification, the results of which suggest that levels of damage have not generally been large. However, the large heave movements which can occur close to trees following their removal are emphasized. Recently published information on the range of influence of various tree species in clays in urban conditions is reviewed and a hierarchy of damaging trees is presented. Some consideration is given to the difficulty of detecting desiccation in clays, with some practical though limited guidance being suggested. Three short case histories are presented to illustrate the influence of trees on both existing buildings and new construction.

Cet article decrit les propriétés de changement de volume des principaux sols argileux britanniques. Les variations saisonnières dans l'humidité du sol sont analysées, ainsi que les reductions importantes que des arbres peuvent produire. L'effet de la sécheresse en Grande-Bretagne pendant les années 1975–76 est évalué selon une classification des dégâts causés aux bâtiments dans beaucoup de cas, mais la conclusion qui en est tirée est qu'en général les dégâts n'ont pas été trop sérieux. Cependant, on souligne les grands mouvements de soulèvement qui peuvent avoir lieu là où des arbres ont été éloignés. On passe en revue des détails récemment publiés au sujet de l'influence exercée par de différents genres d'arbres en ville et on présente une liste des arbres qui causent des dégâts, classés selon leur importance. On discute brièvement la difficulté de détecter la dessiccation dans les argiles, et on suggère des méthodes utiles, bien limitées. Trois exemples sont décrits pour illustrer l'influence exercée par les arbres sur des bâtiments existants et des constructions nouvelles.

INTRODUCTION

Historically, damage in lightweight masonry buildings in Britain has mainly resulted from clay shrinkage beneath shallow foundations undermined by tree roots. However, the damaging effects of clay expansion are becoming increasingly recognized as more use is made of building land recently cleared of large deciduous trees.

Britain's mild, damp climate is very different from those of countries in the humid tropics and the arid zones where most expansive soils occur. For this reason the seasonal soil moisture variations which bring about volume changes are both relatively small and confined to shallow depths. Britain does not therefore experience the problem of substantial ground heave following the covering of a deep profile of desiccated clay which affects construction in the USA, South Africa, Australia, Israel, India and many other countries.

The phenomenon of soil volume changes arising from seasonal moisture changes in clays has been known in Britain for many years; indeed, some 140 years ago Bartholomew (1841) was aware of the problem when he wrote '. . . but in open country situations, during drought, it (strong clay) is apt to split and cause fracture to the building unless the foundation be laid below the range of fissures which occur in it'. More general awareness in the UK was generated by two apparently unrelated events in more recent times. The first, rather surprisingly, was the Second World War. While investigating cases of alleged bomb damage to housing in North London at the end of the War, the Building Research Station identified the combined influence of tree roots, highly shrinkable London clay and a succession of dry summers. The second event was the drought of 1975–76, acknowledged to be the most severe experienced in Britain since records began 250 years ago (Royal Society, 1978). This drought, occurring only four years after the introduction by British insurance companies of indemnity of privately-owned homes against damage arising from subsidence of the land, has had a profound effect on attitudes to house construction, landscape architecture and the existence of cracks in dwellings.

CLAY SOILS

Clays in Britain may broadly be considered in three groups

(*a*) older marine sedimentary deposits of Eocene, Lower Cretaceous and Upper Jurassic vintage
(*b*) glacial clays derived from the last ice age
(*c*) more recent Quaternary deposits of soft, alluvial clays.

* Building Research Establishment.

The surficial glacial clays cover much of eastern, northern and north-western Britain, occurring as far south as the limit of glaciation just to the north of London. Fig. 1 shows the approximate locations of outcroppings of the older clays, the most recent of these being the extensive London clay; also shown are the large areas of shallow boulder clays which become increasingly granular towards the North.

Given a wide variety of depositional conditions and subsequent stress histories, these British clay soils vary considerably in their potential for volume change from the highly plastic Eocene clays (London, Reading) of the South East to the very sandy boulder clays (glacial) of the North. Table 1 (after Avery & Bullock, 1977) shows properties for a range of clays taken from shallow depths. Avery & Bullock found the most expansive smectite (montmorillonite) mineral to exist only in an interstratified molecular form with mica (illite) leading to lower cation exchange capacities (CEC). Weaver & Pollard (1973) suggested that CEC values greater than 10–15 Me/100 g indicate some expansive behaviour. All the soils described in Table 1 possess some volume change potential. However, the CEC values for the well-known, problem clays, such as London, Reading, Gault, Kimmeridge and Oxford, are reduced well below values for the extremely expansive montmorillonitic clays (CEC = 80–150 Me/100 g) by the presence of illite (CEC = 10–40 Me/100 g). In keeping with much international practice (see Holtz & Gibbs, 1957; Williams & Donaldson, 1980) index properties are recommended as a basis for the identification of potentially troublesome clay soils (Building Research Establishment, 1980). Table 2 suggests a classification of potential clay shrinkage, and the potential of some clay soils is shown in Table 3.

Fig. 1. Distribution of stiff clay deposits in Britain

SOIL MOISTURE CHANGES, GROUND MOVEMENTS AND THE INFLUENCE OF TREES

Although rain falls throughout the year in Britain, quantities of rainfall are not particularly high and much of the south and east of the country experiences mean annual rainfalls of less than 750 mm; in the north and west this figure rises to over 1000 mm.

The Meteorological Office publishes data on soil moisture deficit* (SMD) (Meteorological Office, 1979) obtained from evapotranspirational models developed by Penman (1963). Fig. 2 shows a map of the distribution of average peak potential SMD in Britain and demonstrates the coincidence of areas of greatest deficit with areas of highly expansive clays and large conurbations in the South East.

Deficits in the south-east of England usually reach a maximum during September and reduce to zero towards the end of December. Throughout the winter the soil remains at field capacity, except in the vicinity of large trees. The depth below ground with light vegetative cover (grass and shrubs) to which the summer deficit extends has been shown to be about 1·5 m (Ward, 1953).

At the Building Research Station, just north of London on glacial boulder clay (Plasticity Index PI = 38–50%, %<2 μm = 37–51%), values of surface vertical heave of 11 mm and 15 mm were measured in 1977 and 1978 respectively following peak summer deficits of 83 mm and 120 mm of water. These figures suggest a water shrinkage

* When total demand on soil moisture exceeds total replenishment the moisture content decreases in the process of building up a soil moisture deficit.

Table 1. Mineralogy of some British clay soils (after Avery & Bullock, 1977)

Origin	Geology	Sample size	Number of sites	Sample depth below ground level: cm	Clay fraction*: %<2 μm	CEC†: Me/100 g	Ratio of clay minerals‡
Eocene	London clay	13	10	20–95	60 (8·9)	56 (4·7)	1·1 : 2·4 : 6
Eocene	Reading clay	6	3	25–68	58 (10·6)	54 (6·7)	1·3 : 2 : 6·7
Cretaceous	Gault clay	8	6	18–61	57 (4·8)	53 (5·9)	1·25 : 2 : 6·5
Jurassic	Kimmeridge clay	6	5	10–48	63 (12)	48 (10)	1·4 : 2·3 : 5·3
Jurassic	Oxford clay	10	7	13–79	63 (8·2)	55 (11)	1·4 : 2·5 : 6·3
Jurassic	Upper Lias clay	6	3	13–61	61	46	4·5 : 2 : 2·5
Jurassic	Middle Lias clay	6	4	18–71	51 (7·6)	46 (4·1)	3 : 2 : 1§
Jurassic	Lower Lias clay	17	11	10–78	60 (8·4)	50 (5·3)	1·7 : 2·8 : 4§
Triassic	Red Keuper marl	13	11	13–58	52 (10)	38 (7·7)	0 : 6 : 0§
Permian	Permian marl	6	6	13–83	54·5 (12)	35 (7)	1 : 6 : 0§
Carboniferous	Various shales	12	11	23–76	51 (7)	30 (7)	2 : 6 : 0·6§
Devonian	Various sandstone marls and shales	11	9	18–76	43 (5)	31 (8)	0 : 5 : 1·5§

* Standard deviation given in brackets.
† Cation exchange capacity.
‡ Ratios of kaolinite : illite : mica–smectite. Avery & Bullock (1977) did not identify any pure smectite (montmorillonite).
§ Traces of other minerals.

Table 2. Clay shrinkage potential

Plasticity index: %	Clay fraction: %	Shrinkage potential
>35	>95	Very high
22–48	60–95	High
12–32	30–60	Medium
<18	<30	Low

Table 3. Shrinkage potential of some common clays

Clay type	Plasticity index: %	Clay fraction: %	Shrinkage potential
London	28	65	Medium/high
London	52	60	High
Weald	43	62	High
Kimmeridge	53	67	High/very high
Boulder	32	—	Medium
Oxford	41	56	High
Reading	72	—	Very high
Gault	60	59	Very high
Gault	68	69	Very high
Lower Lias	31	—	Medium
Clayey silt	11	19	Low

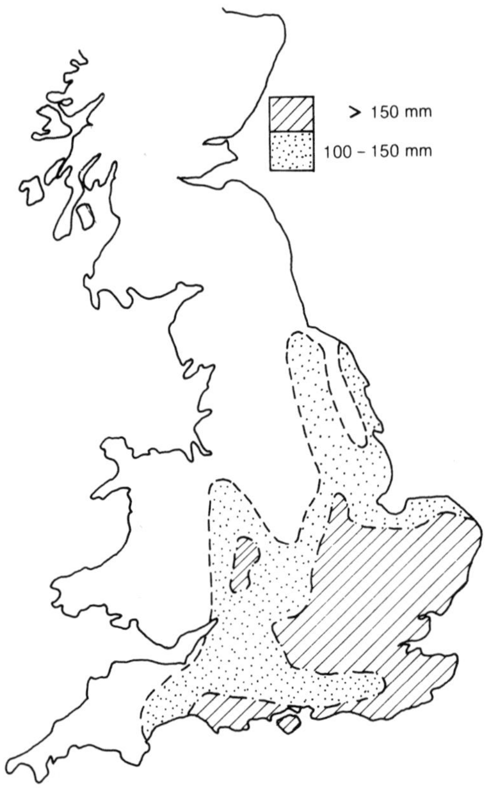

Fig. 2. Distribution of peak summer potential soil moisture deficit (average for 1956–75)

Table 4. Annual number of claims notified to 15 major insurance offices

Year	Number of claims
1971	164
1972	509
1973	817
1974	1705
1975	3821
1976	20 922
1977	11 251
1978	7795
1979	7454

factor, $\Delta w/\Delta z$ (Ward, 1953), of about 8, where Δw is the summer SMD and Δz is vertical ground movement. Ward measured water shrinkage factors of about 4 for various sites on highly expansive clays. Assuming a minimum of 3 for $\Delta w/\Delta z$ would give a conservative estimate of ground surface movement of about 50 mm seasonally for the south-east of England. Ward (1953) measured surface movements at three different sites ranging from 41 to 58 mm and surface heave in excess of 35 mm was recorded at the Building Research Station in the winter of 1976.

Measured water shrinkage factors greater than 3 reflect the heterogeneity and anisotropy induced by particle size variations and predominantly vertical cracking of the ground in summer. The substantial effect that trees can have in increasing moisture deficits in clay soils was demonstrated in Britain some 35 years ago (Ward, 1947). Moisture content profiles in the soil both close to and distant from elm and poplar trees established that desiccation had occurred close to the trees; the differences between these profiles corresponded to an SMD of about 300 mm, suggesting a maximum potential surface movement of 75–100 mm. Further observations by Ward (1953) concerned the distance of trees (mostly elms and poplars) from buildings which had been damaged by clay shrinkage beneath foundations undermined by tree roots. Ward concluded that, on the very highly shrinkable clays, trees which transpired large quantities of moisture in summer could damage buildings at distances up to the height of the tree. An alternative to restricting the relative locations of trees and new housing on these clays was offered in the form of a small diameter pile-and-beam foundation (Ward & Green, 1952). Despite development work on suitable mobile augers to drill pile holes and a limited demonstration of the economic acceptability of such a foundation, the pile-and-beam foundation was not widely adopted for housing.

The reverse of the tree-induced subsidence problem, ground heave following tree removal, was not fully documented in Britain until Samuels & Cheney (1975) presented the results of an investigation of damage to a terrace of single-storey dwellings on 1·2 m deep strip foundations in London clay (PI = 45–60%; %<2 μm = 60–70%). Large elm trees had been felled prior to construction and heave of one end of the terrace was observed to continue for approximately 20 years, with total movements estimated to have been about 100 mm. Desiccation close to the sites of the trees was detected to a depth of about 5 m. Swelling was predicted both from assuming that a desiccated soil moisture profile close to tree positions attains an unaffected profile when water is made available and from oedometer swelling tests. In these tests swelling pressures were first measured at constant volume and then progressively reduced, allowing the samples to swell to equilibrium under total overburden pressure. These methods gave heave predictions ranging from 65 to 70 mm.

Further investigation of heave was carried out on another site at the Building Research Station by Cheney & Burford (1975). Over a period from 1959 to 1980 they monitored the movements of a three-storey office block built on 1·3 m deep strip foundations on boulder clay (PI = 22–29%) immediately after large oak trees had been felled. Level readings were begun four months after tree felling and a maximum heave of 22 mm was recorded at one location in the building eight years after felling. The heave movements caused minor cracking in the building. No predictions of heave were made.

THE 1975–76 DROUGHT

Through the winter of 1975 and the summer of 1976 a severe drought was experienced in northern Europe and hydrological conditions in Britain were the worst experienced since records began. Studies of the effects of the drought (Institute of British Geographers, 1980) show that the south-east of England suffered most severely in terms of soil moisture deficit. The effect in the heavily populated suburbs of London and the Home Counties was dramatic, leading to a substantial increase in property damage claims on insurance. Table 4 shows the growth in claims made on 15 major insurance companies; the effect of the drought can be seen plainly. It has been estimated (Reece, 1980) that the total cost of subsidence damage to dwellings for the period 1971–80 was £250 million, the majority of which resulted from the drought. About 34% of this cost has been attributed to soil movements caused by trees, with the remainder attributable to slight movement of

Table 5. Classification of visible damage to walls with particular reference to ease of repair of plaster and brickwork or masonry (after Building Research Establishment, 1981)

Category of damage	Degree* of damage	Description of typical damage†	Approximate crack width‡: mm
0	Negligible	Hairline cracks of less than about 0·1 mm width are classed as negligible	Up to 0·1
1	Very slight	*Fine cracks which can easily be treated during normal decoration.* Perhaps isolated slight fracturing in building. Cracks rarely visible in external brickwork	Up to 1
2	Slight	*Cracks easily filled. Redecoration probably required. Recurrent cracks can be masked by suitable linings.* Cracks not necessarily visible externally; *some external repointing may be required to ensure weathertightness.* Doors and windows may stick slightly	Up to 5
3	Moderate	*The cracks require some opening up and can be patched by a mason. Repointing of external brickwork and possibly a small amount of brickwork to be replaced.* Doors and windows sticking. Service pipes may fracture. Weathertightness often impaired	5–15 (or a number of cracks up to 3)
4	Severe	*Extensive repair work involving breaking-out and replacing sections of walls, especially over doors and windows.* Window and door frames distorted, floor sloping noticeably§. Walls leaning§ or bulging noticeably, some loss of bearing in beams. Service pipes disrupted	15–25 (but also depends on number of cracks)
5	Very severe	*This requires a major repair job involving partial or complete rebuilding.* Beams lose bearing, walls lean badly and require shoring. Windows broken with distortion. Danger of instability	Usually greater than 25 (but depends on number of cracks)

* In assessing the degree of damage account must be taken of the location in the building or structure where it occurs, and also of the function of the building or structure.
† Ease of repair in italics.
‡ Crack width is one factor in assessing category of damage and should not be used on its own as direct measure of it.
§ Local deviation of slope, from the horizontal or vertical, of more than 1/100 will normally be clearly visible. Overall deviations in excess of 1/150 are undesirable.

foundations constructed prior to the minimum depth requirement of 900 mm introduced in the 1950s.

It is important to be aware of the levels of damage involved. Since the drought the Author has examined over 200 low-rise masonry buildings which had suffered damage allegedly through foundation movement. In doing this, a classification of damage (Building Research Establishment, 1981) was adopted as a basis for comparison. The classification, which uses ease of repair as the main measure of severity, but also takes account of the nature, location and magnitude of cracking, distortion and loss of function, is reproduced in Table 5. Of the many cases of damage examined about 75% fell within and below Category 2, 20% in Category 3 and about 5% in Categories 4 and 5.

The average repair bill for a case of subsidence damage has been estimated (Reece, 1980) at about £4000 at 1980-prices. This high figure reflects the prevalence on the part of those investigating the cases (frequently not engineers and rarely geotechnical specialists) to specify underpinning of foundations. In many cases little or no attempt was made to determine causes of damage or details of the ground conditions. An attitude appears to have built up among owners and those involved in the buying, selling and insuring of houses damaged by

Table 6. Tree species damage ranking

Ranking*	Species	Max. tree height: m	Max. distance recorded: m	Max. distance for 75% of cases: m	Max. distance for 90% of cases: m
1	Oak	16–23	30	13	18
2	Poplar	25	30	15	20
3	Lime	16–24	20	8	11
4	Common ash	23	21	10	13
5	Plane	25–30	15	7·5	10
6	Willow	15	40	11	18
7	Elm	20–25	25	12	19
8	Hawthorn	10	11·5	7	9
9	Maple/sycamore	17–24	20	9	12
10	Cherry/plum	8	11	6	7·5
11	Beech	20	15	9	11
12	Birch	12–14	10	7	8
13	White beam/rowan	8–12	11	9·5	11
14	*Cupressus macrocarpa*	18–25	20	3·5	5

* The ranking has been achieved by weighting the number of species-cases recorded by the ratio of species percentage of the sample to estimated species percentage of the total relevant tree population.

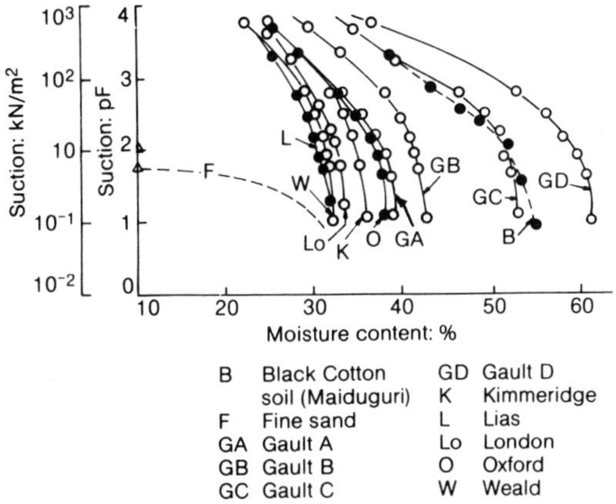

Fig. 3. Relationship between suction and moisture content for undisturbed, expansive clay soils (after Croney, 1977)

Table 7. Index properties for expansive clays of Fig. 3

Clay	Plastic limit: %	Liquid limit: %	Plasticity index: %	Moisture content w	
				pF = 2	pF = 3
Lias	24	60	36	30	26
Weald	25	68	43	31	27
London	26	78	52	32	28
Kimmeridge	24	77	53	34	31
Oxford	24	72	48	36	31
Gault A	25	70	45	37	31
Gault B	29	81	52	41	36
Gault C	34	102	68	51	43
Cotton soil	38	82	44	51	41
Gault D	32	121	89	59	50

foundation movement, in which underpinning, which is almost always applied only to that part of the foundation which is felt to have moved, is regarded as a panacea. Unfortunately, in addition to the monetary effects of the drought, which have been reflected in higher insurance premiums, an increased public awareness of the possible contribution of tree roots to foundation movement and house damage has resulted in many trees being felled without evidence of their culpability. Many underpinning contractors stipulate tree removal as a condition of their guarantee.

TREES IN RELATION TO BUILDINGS

In respect of new construction on clay soils, much argument has arisen between builders, regulatory authorities, planners and landscape architects about the precise relationship between tree roots and ground movements. The so-called '1H' rule deriving from Ward (1953), in which an effective root spread equal to the height of the tree should be assumed, is considered to be aesthetically unacceptable (Flora, 1978), and there has been a reluctance by builders to install deeper foundations. Additionally allegations of damage to existing buildings from root action of trees growing on neighbouring land have generated many cases of litigation.

Recent attempts to produce more information on tree root behaviour and the extent of the zone of influence of trees in clays have produced only limited results. Reynolds (1979) conducted a review of world-wide literature on tree roots but was unable to improve on Ward's (1953) '1H' conclusion. Reynolds recommended research to study moisture content changes and ground movement at various distances from a selection of tree species on various clay sites (see Biddle, 1983).

In another study, Cutler & Richardson (1981) analysed data collected at the Royal Botanic Gardens, Kew, over the period 1971–79. The data, which were obtained from questionnaires sent to customers who had submitted tree roots for identification, yielded information on the distance from the tree to the damaged building, tree height, soil description, root size and other less important factors. Unfortunately, no information on depth of rooting or degree of damage sustained was obtained. The results of the analysis are given for each tree species as the percentage of damage cases recorded for varying distances between tree and damaged building.

Table 6 has been taken from these results and shows a ranking of species in descending order of the incidence of the type of tree in damage cases. The table also shows the likely maximum height of each species for urban conditions on clay soil, the maximum recorded distance at which damage was attributed to tree root action and the distances within which 75 and 90% of cases occurred. It is debatable whether the ranking should be on the basis chosen, rather than, say, on the distance within which 75 or 90% of cases occurred. Various alternative tree rankings in descending order of damaging influence have been compared by Biddle (1979). Cutler & Richardson's (1981) analysis is based on 2600 data cards which makes their sample by far the largest. It could be argued that to preclude 75% of cases of damage caused by tree roots is more realistic than total preclusion. Table 6 suggests this could be achieved by specifying a minimum distance between tree and building of about half the expected tree height. This specification works well with the exception of rowan/white beam, cherry/plum and willow. It is evident from the data that Ward's (1953) original rule of thumb is appropriate if all cases are to be excluded, except for willow and poplar.

THE DETECTION OF DESICCATION

In view of the variability of clay behaviour, differences between hydrological conditions of individual sites and differences in tree species behaviour, it is difficult to offer comprehensive guidance on the likelihood of future damaging tree root action, and the extent and degree of existing root action without recourse to soil tests which are, at least in the house construction industry, outside normal site investigation practice. In the case of existing tree influence the problem becomes one of trying to determine the magnitude of any desiccation from index properties and moisture content values.

Croney (1977) has published curves of moisture content (w) against suction for some undisturbed samples of the more expansive British clays; these are reproduced in Fig. 3 with a curve for fine sand added for contrast. Table 7 gives properties for these soils. Desiccation could be presumed to commence where the rate of change of w (and hence volume) with increasing suction increases noticeably: this point could be taken to be approximately at a suction of pF* = 2 (10 kPa). Similarly, significant desiccation could be taken to have occurred when the suction is sufficiently large that on its release a low-rise building could be lifted; typically, this suction would have a value of pF = 3 (about 100 kPa). A plot of values of w at these suctions, taken from Fig. 3, against Liquid Limit (LL) is shown in Fig. 4; LL is preferred to PI or Plastic Limit (PL) since PL (and hence PI) values depend on a less reliable test procedure than

*The term pF is generally used for suctions and is the $\log_{10}$ of the equivalent height of water measured in cm.

t[illegible]owing r[illegible]

[illegible] (1)

a[illegible]

[illegible] (2)

[illegible] moisture [illegible]and when it

[illegible]mate, method [illegible] to relate it to [illegible]as, Weald and [illegible] for pF = 3 is [illegible]ationship clearly [illegible] and becomes less

[illegible]nducted as part of [illegible] is required in the [illegible]ost cases trees will [illegible]e from the building [illegible]ertheless sufficiently [illegible] influence doubtful. [illegible] to demonstrate the [illegible]t that they exist does [illegible] desiccation. Indeed, [illegible]ed that sections of tree [illegible]rmant until conditions [illegible]re supply activate these [illegible], the measurement of [illegible] samples need not indicate [illegible]servation arises from the fact [illegible] stiff, plastic British clays are [illegible] and possess K_0 values sub- [illegible]er than 1 near the ground surface. [illegible] distribution of total vertical stress [illegible]n below ground level; also shown is a [illegible] porewater pressure distribution (u) below a water table at 2 m from the ground surface. Suppose an oedometer sample is taken from a depth of 1 m below the water table, so that no desiccation can exist in the sample. For the soil properties listed in Fig. 5, at sample depth

$$\sigma'_v = 60 - 10 = 50\,\text{kPa}$$

$$\sigma'_H = 2 \times \sigma'_v = 100\,\text{kPa}$$

where σ'_H is the horizontal effective stress. On sampling, the suction (u_s) set up in the sample due to stress release will be approximately equal to or more likely just less than $(\sigma'_v + 2\sigma'_H)/3$ and will dissipate with time.

However, as an initial approximation it is convenient to take

$$u_s \simeq (\sigma'_v + 2\sigma'_H)/3 = (50 + 200)/3 \simeq 83\,\text{kPa}.$$

If the oedometer swelling test is conducted by applying the total overburden σ_v (= 60 kPa) then swelling could occur because u_s might exceed σ_v, despite the presence of an in situ, positive pore pressure. As is suggested in one of the case histories that follow, Case 3, the detection of swelling in tests of this nature need not indicate desiccation.

CASE HISTORIES

The following case histories offer brief information on some of the practical problems encountered by the Author in dealing with expansive soils and trees. In all the examples neither resources nor time had been available for more thorough examinations of the soil conditions.

Case One

Damage occurred to a four-storey building of

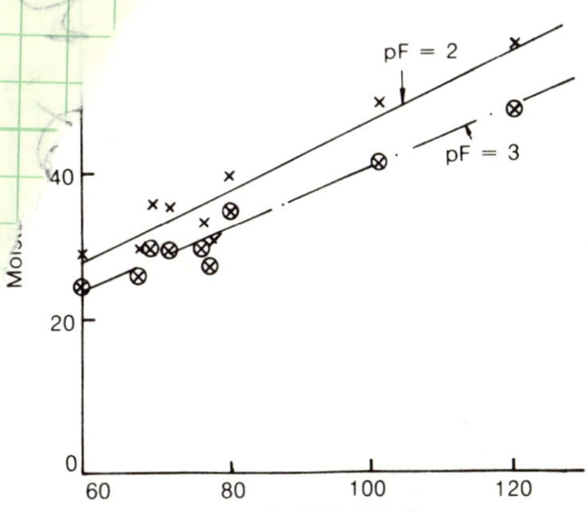

Fig. 4. Moisture content against liquid limit

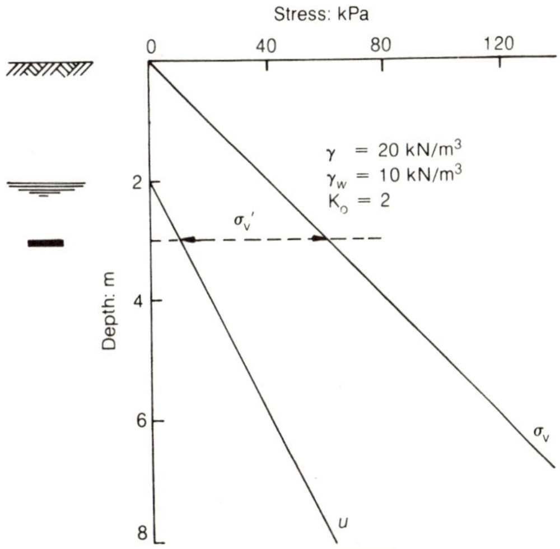

Fig. 5. Distribution of total vertical stress and porewater pressure with depth below ground level

load-bearing brickwork erected at the end of the 19th century. Fig. 6 shows the plan of the front of the triangular-shaped building and the locations of fractures in internal load-bearing walls, floors and ceilings. These fractures, which were first noticed in the mid-1970s, generally increase in width from ground to top floor and indicate rotation of the front portion of the building. No signs of damage were observed in external walls although distortion was evident in some of the numerous window openings; the overall level of damage was classified as Category 3 (Table 5). Preliminary site investigation had consisted of a borehole and two trial pits (excavated in the spring of 1979) at locations shown in Fig. 6. The borehole revealed 2·5 m of made-ground overlying firm to stiff brown fissured London clay, while the trial pits exposed foundations of corbelled brick on a 700 mm wide concrete strip 1·5 m below ground level bearing on brown, sandy clay. The position of a water table was not established. The local geological map indicated a metre of gravel over London clay, broadly confirming these findings.

In the summer of 1979, because the obvious but slow widening of cracks was beginning to raise concern about the long-term stability of the staircase shown in Fig. 6, monitoring of cracks was initiated. A Demec gauge capable of measuring movements to a resolution of about 10^{-3} mm was used. Width changes in the most significant crack are plotted against time in Fig. 7; a seasonal influence can clearly be seen with the crack widening in summer and narrowing in winter.

A busy main road runs parallel to the west wall of the building, separated from it by a wide pavement in which five London plane (*Platanus hispanica*) trees are growing. The trees are approximately 15 m high and are 5–6 m from the west wall. Table 6 indicates that plane is a damaging tree and that in this case the building is well within the zone of influence of the trees. Furthermore, it appears that there is potential for substantial additional growth. No traces of tree root had been found in the borehole and trial pits A and B. When two more pits were excavated (C and D in Fig. 6) they revealed made-ground to 1·5 m, with foundations at 1·7 m on sandy, gravelly, brown clay in which numerous live roots of up to 15 mm diameter were growing. These roots disappeared beneath the foundations. The trend of increasing crack width with time is attributable to the increasing moisture demand of the growing trees.

It is proposed to remove some or all of the trees or to reduce their influence by reducing the volumes of their crowns, or to install barriers to cut off the root systems beneath the building. Monitoring is to be continued to assess the effectiveness of the action taken.

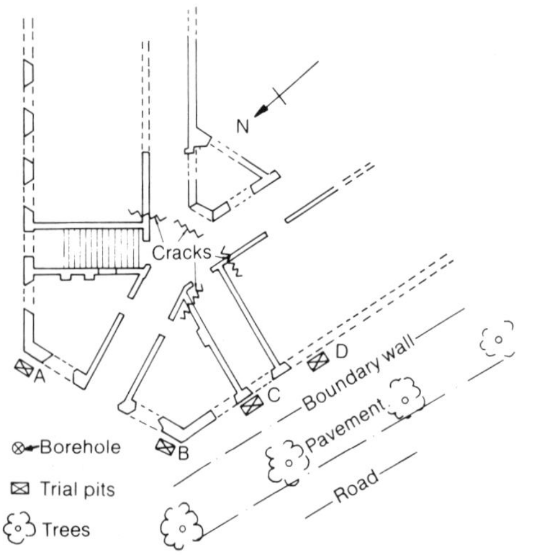

Fig. 6. Plan of damaged building and environs

Case Two

This concerns a private, two-storey dwelling which is situated at the end of a terrace of five similar properties, built in 1972. The form of construction employed is known as cross-wall in which load-bearing walls run transverse to the length of the terrace (Fig. 8). The usual specification of large openings at the front and back of this type of construction makes the structure relatively flexible in both longitudinal bending and torsion. Fig. 9 shows the rear elevation of the dwelling and illustrates the magnitudes of the structural openings in the ground floor.

The home owner first noticed cracks appearing in internal walls, with doors in the rear (south) elevation jamming, in late 1979. In mid-1980 the first of a series of precision levels was taken on bolts inserted near the damp-proof course with

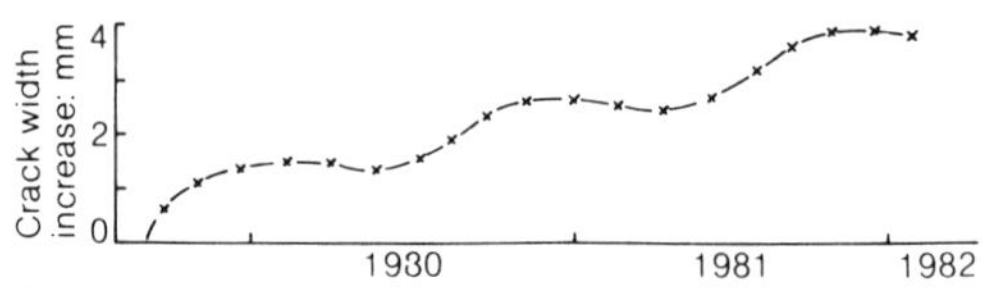

Fig. 7. Crack progression

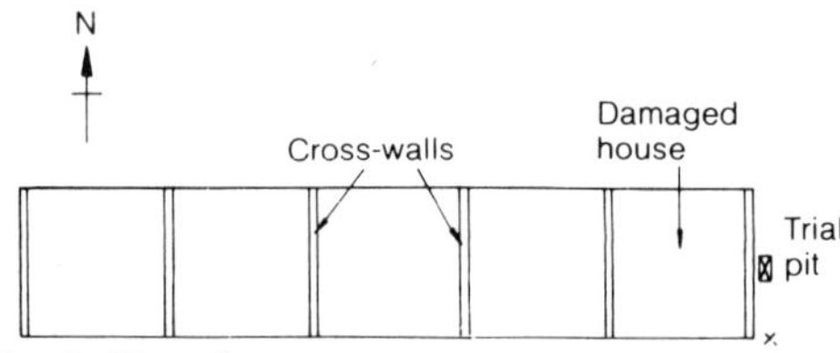

Fig. 8. Plan of terrace

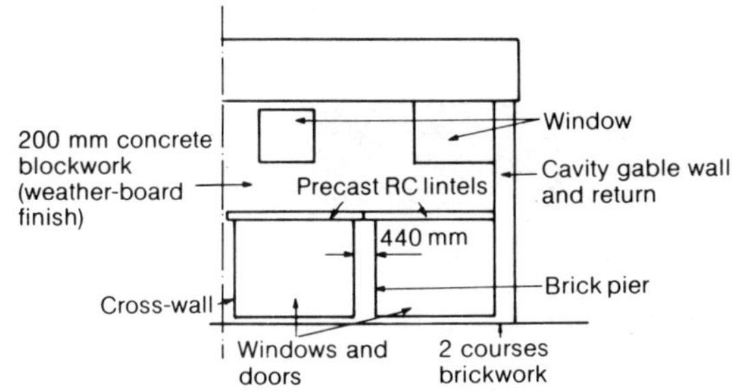

Fig. 9. Rear elevation

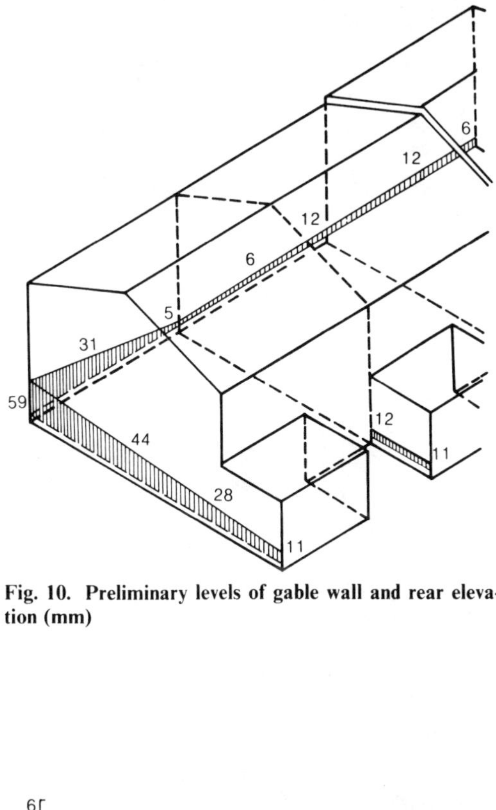

Fig. 10. Preliminary levels of gable wall and rear elevation (mm)

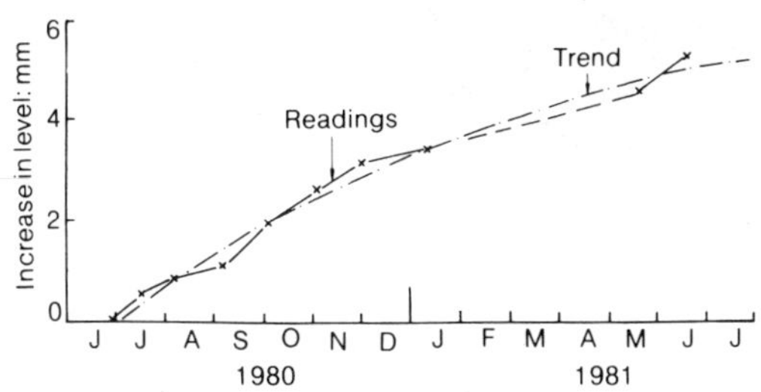

Fig. 11. Heave measurement at point X on Fig. 8

reference to a stable datum installed well away from the building. At the same time a trial pit was dug at the gable wall (Fig. 8) to reveal mass concrete foundations to a depth of 1·25 m in stiff brown, sandy London clay. No soil tests were conducted.

The initial level readings taken on the gable wall and the rear elevation, shown in Fig. 10, suggest an uplift of 59 mm at the south-east corner. Interpretation of initial sets of levels is always imprecise since building tolerances are unknown, although these tolerances tend to be random. The fact that the maximum difference in levels along the entire rear elevation of the other four houses in the terrace amounted to only 20 mm, with a difference of only 6 mm across the west gable, suggests significant heave of the south-east corner of the damaged property. Subsequent level readings at this corner are shown in Fig. 11 and clearly show heaving. No normal seasonal fluctuation is discernible, reflecting the expected foundation stability at 1·25 m below ground level. The trend of the readings suggests that the rate of heave is diminishing, with movements probably terminating in 1982.

The heave is attributed to the removal of a group of trees which aerial photographs reveal to have been growing in the approximate position of the gable wall. With the exception of the damage to the rear elevation, that throughout most of the dwelling has been classed as Category 3. However, the severe distortions which have occurred in the rear elevation, causing large gaps at windows and doors (Fig. 12), the rotation of the brick pier (20 mm out of vertical at the top) and consequential crushing and tension at its base (Fig. 13), and the necessity for reconstruction of the ground floor rear elevation, place the damage in Category 4. Overall slope across the rear elevation measures about 1/100, with local slopes at window heads of 1/70. The stiff cavity gable wall has suffered relatively minor damage (Category 2), with cracks up to 2·5 mm wide in a pattern indicative of hogging distortion. The deflexion ratio Δ/L (Burland & Wroth, 1975) for this wall is estimated to be about $0{\cdot}4 \times 10^{-3}$. Burland & Wroth suggest values of Δ/L, for hogging in load-bearing walls, at which the onset of cracking (expressed as a limiting tensile strain) is reached. For the length-to-height ratio of the gable wall in this example and the hogging mode, the critical value for Δ/L given by Burland & Wroth is about $0{\cdot}25 \times 10^{-3}$. This example suggests that Burland & Wroth's limiting values of Δ/L to prevent cracking are reasonable.

In view of the expected termination of heave, the appropriate solution in this case is to rebuild the rear ground floor elevation and to make-good elsewhere.

Fig. 12. Distortion of right-hand door, rear elevation

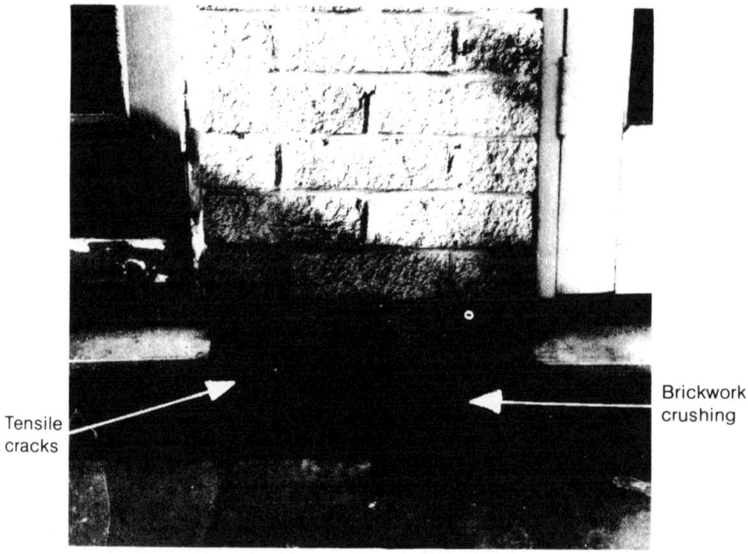

Fig. 13. Rotation of brick pier, rear elevation

Case Three

This case involves the interpretation of data from a site investigation for lightweight, single-storey and two-storey dwellings adjacent to rows of poplar and willow trees of heights up to 20 m. The distances between the trees and the proposed buildings range between 5 and 16 m. Fig. 14 shows a description and properties of samples taken from a borehole (A) 7 m from the trees and another (B) 20 m from the trees. A water table was observed in borehole A at 1·6 m, the borehole having been drilled at the height of summer.

In view of the discovery of fine roots to a depth of 3·5 m in borehole A, the proximity of trees well-known to have caused damage (Table 6) and the measurement of oedometer swelling (Fig. 14), concern was expressed that future tree action or heave following tree removal could cause building damage. Accordingly, a recommendation was made by the site investigation contractor that foundations consisting of 12 m long piles with fully suspended capping beams and floors be constructed.

Dismissing the top two metres of the profile since the soils there are of only intermediate plasticity, the presence of a shallow summer water table in borehole A makes the existence of desiccation unlikely. Furthermore, the in situ moisture content at 2·75 m (borehole A) of 34% is well above both 0·4 LL and PL+2 (LL = 72, PL = 27). At 3·25 m in borehole B, the in situ moisture content of 29% is again above 0·4 LL and PL+2 (LL = 66, PL = 24). The swelling measured in both samples at these depths is attributed either to inadvertent sample drying prior to the test or to the release of large lateral pressures on sampling, as described earlier. It is concluded that the expensive foundation measures proposed are not justified on the data presented. At the time of writing no decision on the foundation design had been taken by the developer.

CONCLUSIONS

Simple, reasonably reliable means exist to identify expansive British clay soils from values of the index properties. Data show that many of these soils, especially in the south-east of England possess large potential for volume change. However, without tree influence, the mild damp British climate ensures that significant summer soil

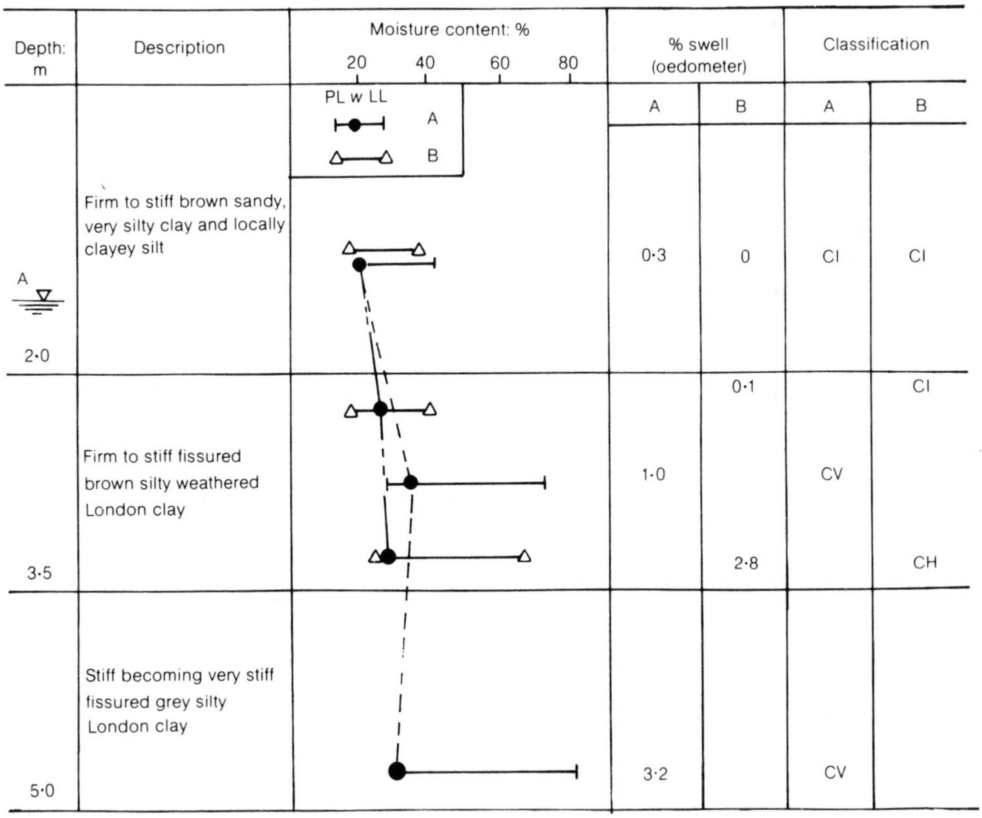

Fig. 14. Ground profile and sample swelling

moisture deficits are confined to the top 1–1·5 m and field capacity is re-established each winter. Deeper, permanent deficits can be created by large trees and there is much uncertainty surrounding the size and extent of these deficits. Crude design rules which either restrict coexistence between trees and buildings to an aesthetically unacceptable degree or else lead to increased foundation costs cause difficulty especially for marginal cases. Enhanced public sensitivity to what are usually low levels of damage arising from tree and building interaction has increased these difficulties. There is some evidence to suggest that 75% of cases of tree-induced damage could be eliminated by separating trees and buildings by distances of about half the tree height.

The principal difficulty both in investigating new clay sites with trees present or removed, and in diagnostic examination of damaged buildings with trees in the vicinity, is in determining soil moisture suction values. There is very little data available on the suction characteristics of British clays. Based on limited data, a tentative means of determining the likelihood of significant desiccation by relating in situ moisture contents to liquid limits is offered. Caution is advised in the interpretation of values of swelling indicated in oedometer tests; swelling of samples taken from shallow depth in over-consolidated clays can result among other things from stress relief on sampling.

ACKNOWLEDGEMENT

The work described was carried out as part of the research programme of the Building Research Establishment of the Department of the Environment. This Paper is published by permission of the Director.

REFERENCES

Avery, B. A. & Bullock, P. (1977). *Mineralogy of clayey soils in relation to soil classification.* Soil Survey Technical Monograph No. 10. Harpenden: Soil Survey.

Bartholomew, A. (1841). *Specifications for practical architecture*, 1st edn. London: J. Williams.

Biddle, P. G. (1979). Tree root damage to buildings: an arboriculturist's experience. *Arb. J.* **3**, No. 6, 397–412.

Biddle, P. G. (1983). Patterns of soil drying and moisture deficit in the vicinity of trees on clay soils. *Géotechnique* **33**, No. 2, 107–126.

Building Research Establishment (1980). *Low-rise building on shrinkable clay soils: part 1.* Digest 240. London: HMSO.

Building Research Establishment (1981). *Assessment of damage in low-rise buildings, with particular reference to progressive foundation movement.* Digest 251. London: HMSO.

Burland, J. B. & Wroth, C. P. (1975). Settlement of buildings and associated damage. *Proc. Conf. Settlement of Structures, Cambridge*, pp. 611–654. London: Pentech Press.

Cheney, J. E. & Burford, D. (1975). Damaging uplift to a three-storey office block constructed on a clay soil following the removal of trees. *Proc. Conf. Settlement of Structures, Cambridge*, pp. 337–343. London: Pentech Press.

Croney, D. (1977). *The design and performance of road pavements*, p. 674. London: HMSO.

Cutler, D. F. & Richardson, I. B. K. (1981). *Tree roots and buildings*, p. 94. London: Construction Press.

Flora, T. (1978). Treeless towns? *J. Inst. Landsc. Archit.* **121**, 10–12.

Holtz, W. G. & Gibbs, J. J. (1957). Engineering properties of expansive clays. *Trans. Am. Soc. Civ. Engrs* **121**, 641–663.

Institute of British Geographers (1980). *Atlas of drought in Britain 1975–1976*, p. 30. London: Institute of British Geographers.

Meteorological Office (1979). *Estimated soil moisture deficit and potential evaporation over Great Britain—explanatory notes*, p. 5. Bracknell: Meteorological Office.

Penman, H. L. (1963). *Vegetation and hydrology.* Technical Communication No. 53, p. 24. Harpenden: Commonwealth Bureau Soils.

Reece, R. A. (1980). Insurance: the cost of mistakes. *Proc. Sem. Trees in relation to Construction*, pp. 28–36. London: British Standards Institution.

Reynolds, E. R. C. (1979). *A report on tree roots and built development*, p. 20. Department of the Environment. London: HMSO.

Royal Society (1978). Scientific aspects of the 1975–76 drought in England and Wales. *Proc. R. Soc.* A, 363.

Samuels, S. G. & Cheney, J. E. (1975). Long-term heave of a building on clay due to tree removal. *Proc. Conf. Settlement of Structures, Cambridge*, pp. 212–220. London: Pentech Press.

Ward, W. H. (1947). The effect of fast growing trees and shrubs on shallow foundations. *J. Inst. Landsc. Archit.* **11**, 7–16.

Ward, W. H. (1953). Soil movement and weather. *Proc. 3rd Int. Conf. Soil Mech., Zurich* **4**, 477–482.

Ward, W. H. & Green, H. (1952). House foundations: the short bored pile. *Proc. Pub. Wks Mun. Services.* pp. 1–16. London: Institution of Civil Engineers.

Weaver, C. E. & Pollard, L. D. (1973). The chemistry of clay minerals. *Developments in sedimentology*, Vol. 15. New York: Elsevier.

Williams, A. A. B. & Donaldson, G. W. (1980). Building on expansive soils in South Africa: 1973–1980. *Proc. 4th Int. Conf. Expansive Soils, Denver* **2**, pp. 834–844. New York: American Society of Civil Engineers.

Patterns of soil drying and moisture deficit in the vicinity of trees on clay soils

P. G. BIDDLE*

Existing guidelines on the proximity of trees and buildings are based on the concept of permitting minimum foundation depths and then trying to ensure sufficient separation between the tree and building to prevent damage. Such guidelines are impractical and are not universally accepted or implemented. The alternative approach is to increase foundation depth to permit most tree planting designs and natural regeneration, but in the past very little information has been available on the necessary depths. The Paper describes research into the patterns of deficit near 36 different trees, covering a range of tree species and clay types. It demonstrates that poplars have much greater effects than the other species which are investigated. Clay type makes little difference to the patterns of deficit, although it is important because of differences in permeability and the amount of shrinkage which occurs. Differences between individuals of the same tree species are comparatively small, unless there are anomalies in the soil conditions. It is concluded that a modest increase in foundation depth to about 1·5 m would permit most planting designs accommodating the majority of tree species (except poplar, elm, oak and willow) on soils of very high shrinkage. Lesser depths could be used for soils of lower shrinkage potential. Special precautions need then only be taken with species of high water demand.

Les règlements concernant la proximité des arbres et des bâtiments sont essentiellement basés sur des profondeurs minimales permises des fondations, combinées avec des distances entre les arbres et les bâtiments qui sont suffisantes pour empêcher les dégâts. De tels règlements sont peu pratiques et ne sont pas adoptés universellement. Dans une autre approche, on peut augmenter la profondeur des fondations pour permettre une croissance raisonnable des arbres. Cependant, jusqu'à present il n'y avait pas de données en ce qui concerne les profondeurs nécessaires. Cet article présente des études du déficit d'humidité dans le voisinage de 36 arbres d'essences différentes sur de divers types d'argile. On démontre que les peupliers produisent des effets beaucoup plus marqués que les autres essences examinées. Le type d'argile exerce très peu d'influence sur le déficit d'humidité mais peut avoir de l'importance à cause des différences de perméabilité et de retrait possibles. Il n'y a que relativement peu de différences entre les arbres individuels de la même essence, à moins que les conditions du sol ne soient anomales. En conclusion une augmentation légère de la profondeur des fondations (jusqu'à 1·5 m environ) permettrait la plantation d'arbres de la plupart des essences (à l'exception du peuplier, de l'orme, du chêne et du saule) sur des sols susceptibles d'un retrait très élevés. De moindres profondeurs pourraient être adoptées pour des sols à faible potentiel de retrait. Dans ce cas des précautions spéciales ne doivent être prises que pour des essences qui exigent une grande quantité d'eau.

INTRODUCTION

The first guidelines on the safe proximity of trees and buildings on clay soils in Britain (Building Research Station, 1949) advocated the separation of trees and buildings to distances equal to the height of the tree, or 1·5 times the height for rows or groups of trees. This guideline was later refined (National House-Building Council, 1969) to distinguish greater effects from poplar, elm and willow trees, and to recommend increased foundation depth for new houses where existing trees were closer than the minimum distance.

The practicality of such recommendations is dubious, as it would virtually eliminate all trees from urban development, and was considered unacceptable by the public who enjoy the aesthetic benefits of tree-lined streets and gardens. However, the severe drought of 1975–76 refocussed attention on the guidelines, and has led to widespread pressure for the felling of trees, and fierce debate between the tree and building lobbies.

As a result of these problems, the Department of the Environment commissioned a report to evaluate the current state of knowledge and the appropriateness of the existing recommendations (Reynolds, 1979). Among the conclusions in this report were

'(*b*) it is an inappropriate solution to the problem to rely solely on regulating the trees' proximity to the building

(*c*) the foundation is the best understood part of the system and the most adaptable through which to reduce the risk of damage to buildings

(*e*) very little progress in regulating urban trees on shrinkable clays can be made without additional information from critical field observations'.

The observation that the policy of trying to

* Tree Conservation Ltd, Wantage, Oxon.

separate trees and buildings may be the inappropriate solution is also seen in various new guidelines. The British Standard Code of Practice BS 5837 (British Standards Institution, 1980) officially recognized for the first time that

> 'most buildings are likely to come into close proximity with trees during their useful life, owing to self-sown trees or planting'.

This Standard also records

> 'Where the presence of shrinkable clay has been established and where roots have been found, or are anticipated in the future, it is advisable to take precautionary measures in the design of the building's foundations'.

Digest 240 (Building Research Establishment, 1980a) notes that

> 'The design of foundations for low-rise buildings must take account of the shrinkage and swelling behaviour of clays, as the moisture content varies according to climatic and vegetation conditions. In practice two courses of action are open: firstly, to estimate the potential to swell and shrink and then to try to avoid events which cause changes in moisture content; or secondly, to accept that swelling or shrinkage will occur and to take account of this in design. The foundations can be designed to resist resulting ground movements, or the superstructures designed to accommodate movement without damage'.

Digest 241 (Building Research Establishment, 1980b) notes that

> 'Buildings can be safely sited close to trees, and vice versa, on clay soil of very high shrinkage potential provided adequate foundations are used. It is necessary to take the foundations below the existing or probable zone of desiccation, or root growth as determined by investigation . . .'.

These guidelines and Codes of Practice therefore demonstrate a new philosophy towards preventing subsidence damage. Whereas previously the policy had been to try to ensure sufficient separation between trees and buildings and use minimum foundation depths, the emphasis should now be on accepting that trees will be planted or grow near buildings, and designing the foundations accordingly. Safety should be achieved by depth of foundation, rather than by lateral separation.

However, it was also recognized by Reynolds (1979) that little progress can be made without additional information from critical field studies. Some information came from the study of tree root spread (Cutler & Richardson, 1981), but their observations on the proximity of trees to buildings are based on the existing housing stock, much of which is on very shallow foundations. The study provides no data on the depth of effect of trees.

As a result of this lack of information, Tree Conservation Ltd has been commissioned to carry out two similar research projects, involving detailed investigations of the pattern of soil moisture change in the vicinity of different trees on different clay soils. The first study, commissioned in 1979 by Milton Keynes Development Corporation (MKDC), looks at three tree species, poplar (*Populus x euramericana*), lime (*Tilia x europaea*) and horse chestnut (*Aesculus hippocastanum*), growing on the two main soil types which outcrop in the Milton Keynes area (Oxford and boulder clay). There are two replicates of each tree species on each soil, involving a total of 12 trees. This study was required because of the massive tree planting programme which was being undertaken within this New Town development, with much of this planting being within the potential mature heights of the trees from the buildings. Concern was being expressed that this planting might cause damage in future.

The second project was commissioned by the National House-Building Council (NHBC). Their Practice Note 3 (National House-Building Council, 1969) provides the only guidelines on the increased foundation depths which are required when houses are being built closer to trees than the tree height. In the absence of any other recognized guidelines, this Practice Note is now being used in situations well beyond its original intended purposes. However, it is recognized that this Note gives inadequate emphasis to the variation in properties of clay soils and the difference between tree species, and a revision of it is in preparation. This revision is based on the existing state of knowledge, but the NHBC recognize that further study will be needed to enable this to be more accurately revised in future.

The study for the NHBC involves an additional 24 trees (making 36 in all). These are located on a variety of clay soils, chosen to try to reflect the range of clay shrinkage potentials as defined in Digest 240 (Building Research Establishment, 1980a). Three broad leaved species of tree are being studied to reflect the variation in water demand which is believed to exist between different species. These species are poplar (high water demand), horse chestnut (moderate water demand) and silver birch *Betula pendula* (low). In addition one conifer, Leyland cypress (*x Cupressocyparis leylandii*), has been included. This species is a hybrid, and experience in the field has shown that one of its parents, Monterey cypress, can have an exceptionally severe desiccating effect on clay soils.

The project on the initial 12 trees for MKDC has now been underway for 4 years, and the 24 trees for NHBC for 2 years. This has provided a mass of data which are summarized in this Paper. However, it is considered that the full benefits of this project can only be realized when there is a drought of reasonable severity, so that an assessment can be made of the additional soil drying which develops under such conditions. The projects will continue to run until that time, and it is hoped that then new information on the full effects of trees on clay soils during drought conditions will be obtained.

MATERIALS AND METHODS

The patterns of seasonal change of soil moisture content have been monitored using a series of neutron probe boreholes in the vicinity of the trees.

Neutron probe

The instrument used is a model 225 Wallingford soil moisture probe, incorporating a 70 millicurie americium-241/beryllium source. The probe can be lowered down access tubes which are permanently inserted in the ground, and this provides a measure of the moisture content of the surrounding soil. Repeated measurements can be made in these access tubes at any appropriate time interval, thus providing measurement of the seasonal or long-term changes in moisture content. The access holes cause the minimum disruption to the soil or roots, and if carefully inserted do not affect the natural drainage patterns of the soil.

The tubes are 44·5 mm in diameter, made of 16 g aluminium, and sealed with a cap at the bottom to prevent entry of water. They are hammered into holes of the same diameter which are reamed out of the clay with a combination of manual screw augers and mild steel reamers. The tubes were inserted during August/September when the clay is in its driest condition; any subsequent rewetting and swelling of the clay helps to ensure as tight a fit as possible. Most tubes were inserted to 3 m depth, but longer 4 m tubes were used in any situation where it was suspected that root effects might extend. The tops of the tubes are buried just below ground level, sealed, and covered with the natural vegetation of the area (usually grass) which is lifted aside when readings are taken. Other methods adopted are those described by Bell (1976).

Location of access tubes

Five tubes were inserted for each tree at differing distances, wherever possible along a single radius. The distances from the tree were related to the height of the tree at the time of insertion. For the MKDC project these were 0·1, 0·2, 0·4, 0·8 and a comparison tube at 2 times tree height (referred to as $0{\cdot}1h$, $0{\cdot}2h$ etc. where h is the height of the tree). In the light of experience gained from this work, distances for the NHBC project have been varied depending on the species involved, with particular emphasis near the edge of the main zone of influence. In all cases one of the tubes is located beyond any possible influence of the tree, but otherwise in as similar conditions as possible, to provide a comparison.

The use of tree height as a parameter for assessing the extent of effects of trees has been questioned recently (Reynolds, 1979; Cutler & Richardson, 1981). However, some parameter is required to enable comparison to be made between trees of differing size and their different water requirements. Tree height is the most logical parameter; it is commonly used at present and the results appear to confirm that it is appropriate.

Location of trees

Suitable trees for this research were required to be growing on the appropriate clay, isolated from trees of other species, to have sufficient space and access for instrumentation and to be clear of other factors (e.g. streams, ditches) which might influence the moisture content. These criteria were most commonly found with trees growing in parks or around playing fields.

The work for MKDC required investigation of Oxford and boulder clays which outcrop in that

Table 1. Combination of tree species and soil types

Soil type	Tree species	Number of trees
Gault clay	Poplar	2
	Horse chestnut	3
	Silver birch	1
	Leyland cypress	2
London clay	Poplar	3
	Horse chestnut	4
	Lime	1
	Silver birch	2
	Leyland cypress	2
Oxford clay	Poplar	2
	Horse chestnut	2
	Lime	2
Boulder clay	Poplar	2
	Horse chestnut	2
	Lime	2
	Silver birch	1
	Leyland cypress	2
Clay silt	Poplar	1
		Total 36

area, and NHBC required a range of clay soils to represent clays of different shrinkage potential. It was anticipated that Gault clay around Cambridge would exemplify very high shrinkage clay, and that London clay in the north-west suburbs of London would be typical of high shrinkage potential. Oxford clay also comes within the high shrinkage range. The medium shrinkage soils were the boulder clay soils in the Milton Keynes area. One tree on clay silt (low shrinkage) was used. The combination of tree species and soil types is summarized in Table 1.

Calculation and presentation of results

Readings in each access hole are taken at 10 cm intervals down the length of the tube, ensuring that the probe is at exactly the same depth on each occasion. This sampling interval is increased to 20 cm in the deeper soil layers where there is no evidence of any seasonal changes.

A 64 second count period is used for all measurements. The soil moisture content is calculated from these counts/second results by use of the formula

$$\text{Moisture content } \% = ((0{\cdot}958R/R_w) - 0{\cdot}012)\ 100$$

where R is the count rate in soil and R_w is the count rate in a water standard.

This is the formula recommended by the Institute of Hydrology (Bell, 1976) as typical for the calibration in clay soils. No attempt has been made to allow for variation in the soil type (although it is known that at some positions there is a comparatively sandy soil) nor to provide more precise calibration for any inherent differences between the clay types. It is appreciated that this introduces some error in the measurement of absolute moisture content, but it makes very little difference to the accuracy of measuring change in moisture content.

The moisture content is calculated on a volume/volume* basis, not weight/weight basis as is more commonly used when determining moisture content by oven-drying soil samples. These volumetric measurements have the advantage of permitting direct calculation of the actual volume of moisture in each layer of soil, and by measuring the seasonal changes, the water deficit can be calculated.

Readings on all of the access tubes are taken in early April, just before active water uptake starts for the year. In the MKDC project which has been running for three years, the average of the three readings is used. On some selected trees readings are taken at intervals through the summer, but the main set is taken in September when water deficits are at a maximum before leaf fall and the onset of winter rains. These two sets of readings therefore provide a measure of the soil in its wettest and driest conditions, and the results which are presented concentrate on this comparison.

The results are presented in diagrammatic form in two ways

(*a*) the reduction in soil moisture content in autumn compared with spring. This is shown in Figs 1–10 and 15–20 by the density of shading—the darker the shading the drier the soil

(*b*) contours of soil moisture deficit. These are calculated by multiplying the change in moisture content by the appropriate layer thickness represented by the measurement (in most cases 10 cm). Thus in a 10 cm thickness of soil, a 4·5% reduction in moisture content would produce a 4·5 mm water deficit. These water deficits are added cumulatively from the base of the zone of desiccation to determine the water deficit of each level, and the results for the five holes at each tree are presented as contours of moisture deficit.

All the figures show soil depth (in cm) as the vertical axis. No results are shown in the surface 30 cm, as these are affected by either the vandal-proof cap or by the excavations to expose the ends of the tube. The horizontal axis in Figs 1–10 and 15–20 shows the distance from the tree (the centre of the tree being at the position of the vertical axis). The position of access holes up to distances of 1·5*h* are at the appropriate proportional distance from the tree along this axis, but the distances for the comparative holes at 2*h* or 3*h* are foreshortened. The distance of the tubes from the tree is shown both as the proportion of height, and as the actual distance. In the figures the contours of water deficit, and the density of shading to show reduction in moisture content are interpolated between the different tubes as logically as possible, bearing in mind the descriptions of the soil profiles obtained when the tubes were inserted.

In addition to the results for the individual trees, results have been calculated to show the average for all of the trees of the different species. These combined average results are particularly useful for comparing the effect of different species.

RESULTS

Selected examples of the results from the 36 trees under investigation are used to illustrate different aspects: variation between species, variation between individuals of the same species, variation between soil types, influence of abnormalities in soil profile, influence of soil permeability and permanent moisture deficits, average results for the

* Volume of water per unit volume of soil dried at 105 °C.

Table 2. Summary of conditions at sites shown in Figs 1–22

Figure	Clay type*	Index properties			Location	Tree species	Height: m	Surrounding vegetation
		PL	LL	PI				
1	Boulder	20	43	23	Milton Keynes	Lime	14	Mown grass
2	Boulder	23	46	23	Milton Keynes	Horse chestnut	14	Close mown grass
3	Boulder	18	36	18	Milton Keynes	Poplar	17	Mown grass
4	Boulder	17	38	21	Milton Keynes	Silver birch	8	Mown grass
5	Boulder	19	45	26	Milton Keynes	Leyland cypress	9	Mown grass
6	Boulder	17	36	19	Milton Keynes	Lime	15	Grazed paddock
7	Oxford	37	66	29	Milton Keynes	Horse chestnut	15	Mown grass and herbaceous border
8	London	28	88	60	Wembley	Horse chestnut	14	Mown grass
9	Gault	31	82	51	Cambridge	Horse chestnut	12	Close mown grass
10	Gault	21	52	31	Cambridge	Silver birch	12	Close mown grass
11	Boulder	18	36	18	Milton Keynes	Poplar	17	Mown grass
12	Oxford	26	57	31	Bedford	Poplar	15	Mown grass
13	London	30	71	41	Wembley	Poplar	20	Mown grass
14	Gault	30	73	43	Cambridge	Poplar	19	Close mown grass
21	Boulder	18	36	18	Milton Keynes	Poplar	17	Mown grass
22	Boulder	18	36	18	Milton Keynes	Poplar	17	Mown grass

* Geological formation for the clay strata is London clay: Eocene; Gault; Cretaceous; Oxford clay: Upper Jurassic; Boulder clay: Glacial, composed mainly of Jurassic materials.

Key to Figs 1–10 and 15–20

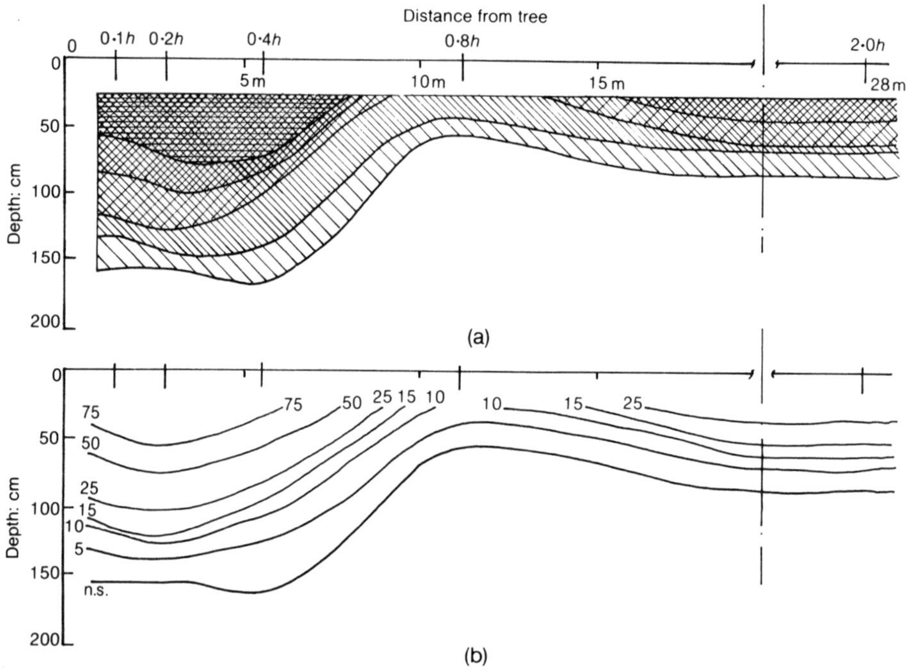

Fig. 1. Lime on boulder clay, 1 September 1982: (a) reduction in soil moisture content; (b) soil moisture deficit (mm)

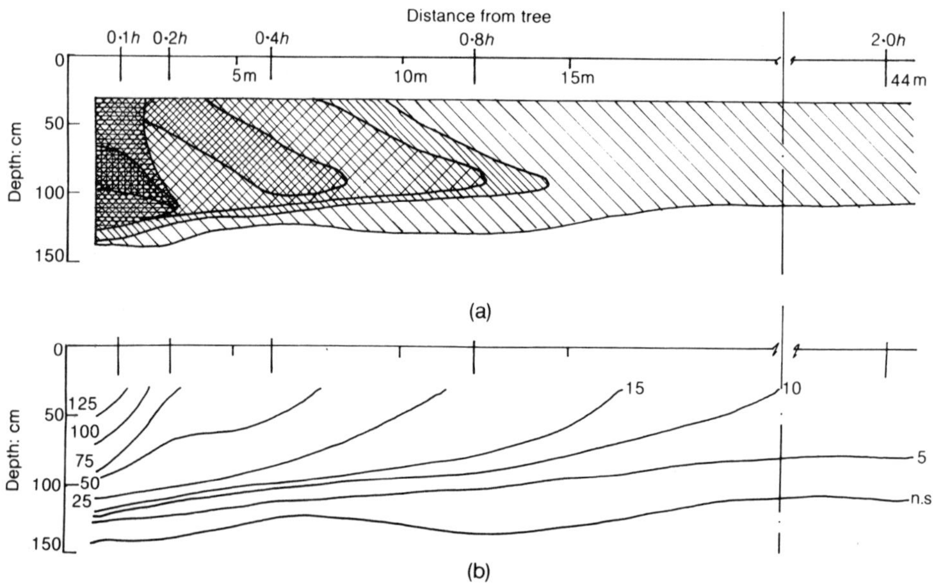

Fig. 2. Horse chestnut on boulder clay, 10 September 1982: (a) reduction in soil moisture content; (b) soil moisture deficit (mm)

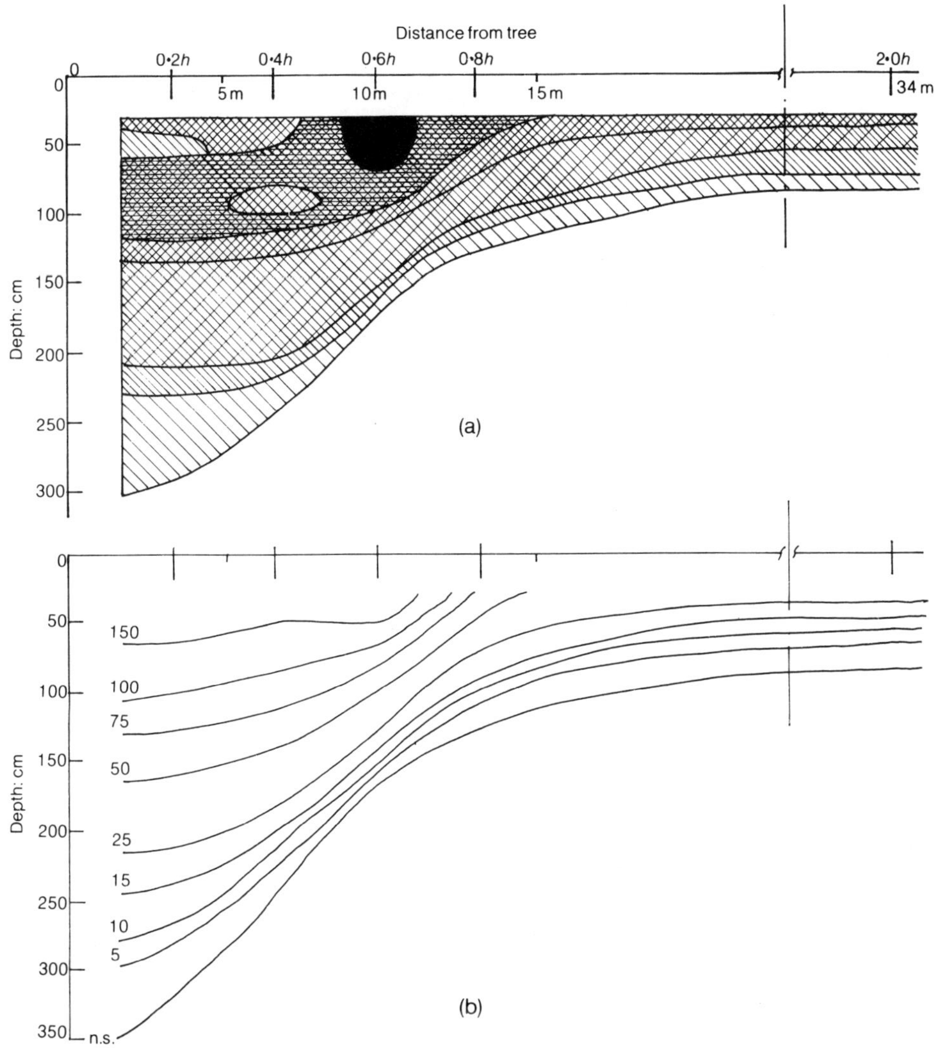

Fig. 3. Poplar on boulder clay, 14 September 1982: (a) reduction in soil moisture content; (b) soil moisture deficit (mm)

different species, development of deficit during the year, and difference of profile between years. The results are presented in Figs 1–22. Details of the sites are summarized in Table 2.

Variation between species

Figures 1–5 present typical results for the five species under investigation, each of these being on boulder clay soil.

Figure 1 (lime) shows effects extending to 160 cm at distances to 0·4*h*, but at 0·8*h* there is less effect than at the comparison position. Close to the tree there is a 25 mm soil moisture deficit at a depth of 100 cm.

Figure 2 (horse chestnut) shows pronounced effects close to the tree, but these do not extend below 130 cm.

By contrast, Fig. 3 (poplar) shows effects to 290 cm at 0·2*h*, to 150 cm at 0·6*h* and shallow effects at 0·8*h*. The soil moisture deficit at 100 cm is just under 100 mm at 0·2*h* and at 0·6*h* there is a 50 mm deficit. This result indicates a far greater effect by poplar, both in depth and lateral extent, than any of the other species under investigation.

Silver birch (Fig. 4) shows some effects to 150 cm at 0·2*h*, but there is no significant difference between the other positions and the comparison hole. Similarly, Leyland cypress (Fig. 5) shows a localized effect, but not extending below 100 cm.

Variation between individuals of the same species

Results between individuals of the same species were generally similar, with any major differences usually being attributable to soil abnormalities (see

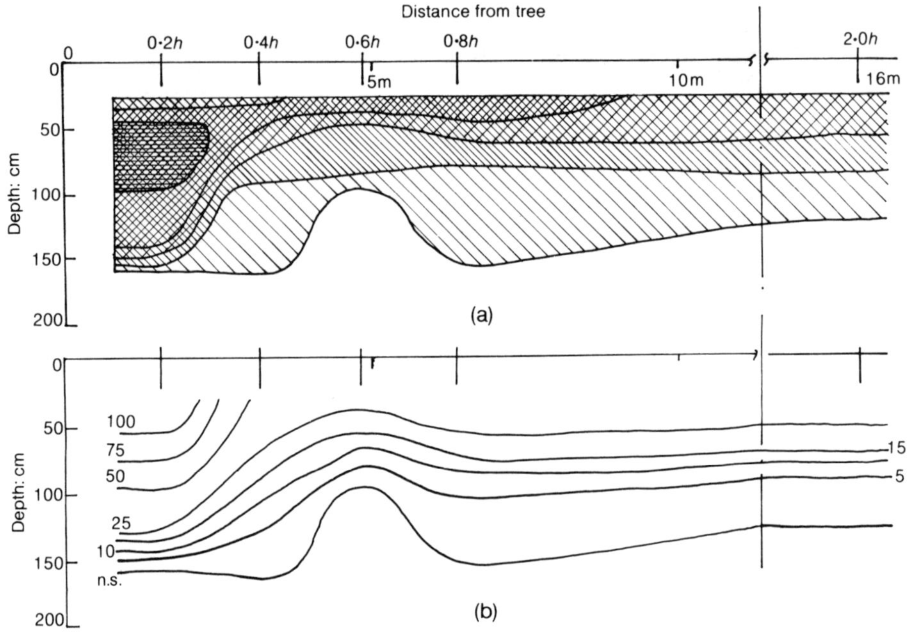

Fig. 4. Silver birch on boulder clay, 13 September 1982: (a) reduction in soil moisture content; (b) soil moisture deficit (mm)

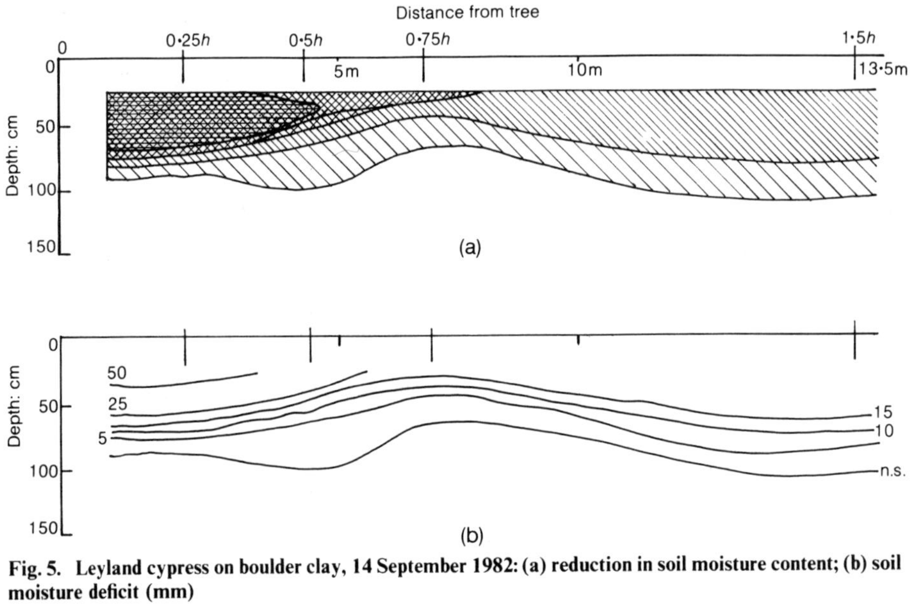

Fig. 5. Leyland cypress on boulder clay, 14 September 1982: (a) reduction in soil moisture content; (b) soil moisture deficit (mm)

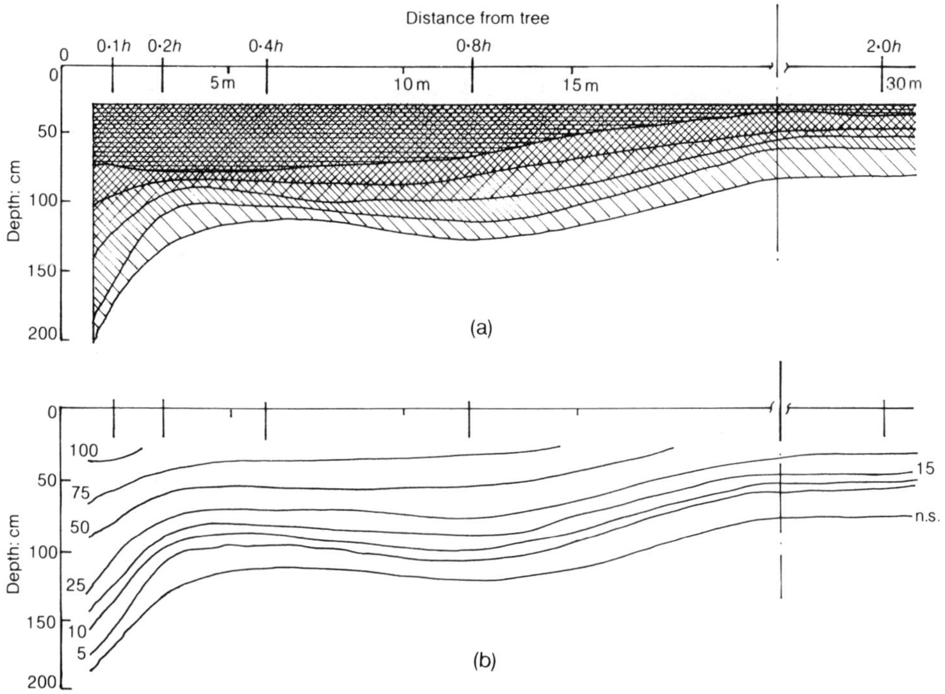

Fig. 6. Lime on boulder clay, 13 September 1982: (a) reduction in soil moisture content; (b) soil moisture deficit (mm)

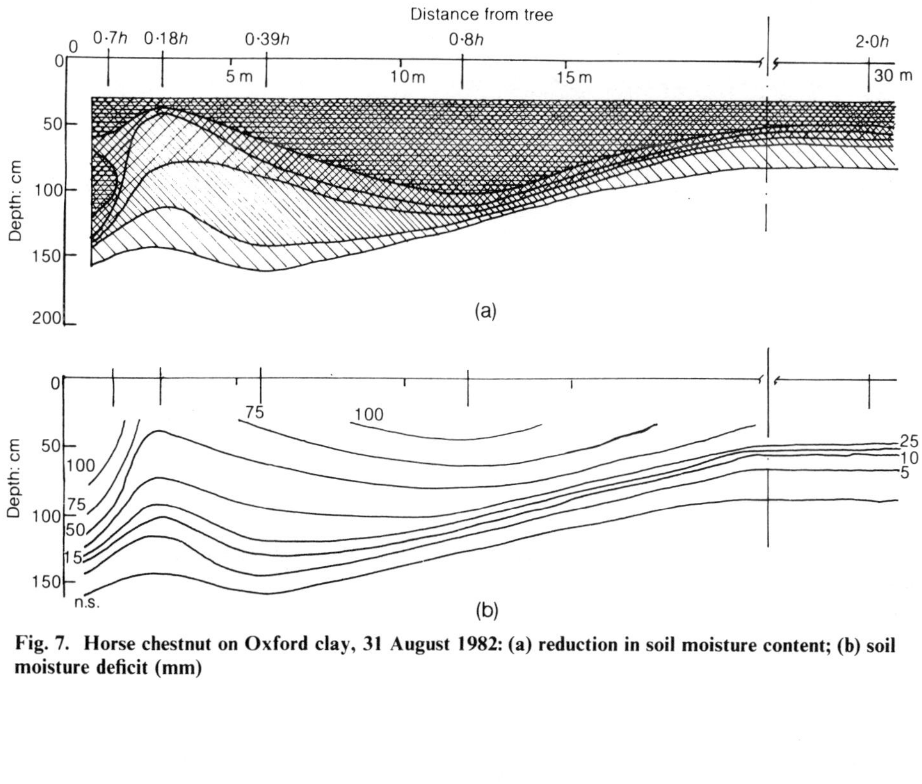

Fig. 7. Horse chestnut on Oxford clay, 31 August 1982: (a) reduction in soil moisture content; (b) soil moisture deficit (mm)

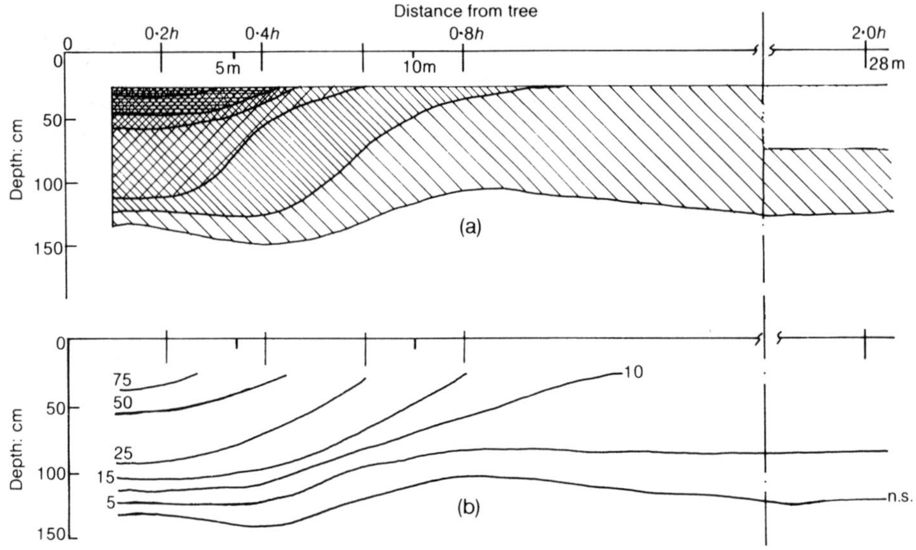

Fig. 8. Horse chestnut on London clay, 29 September 1982: (a) reduction in soil moisture content; (b) soil moisture deficit (mm)

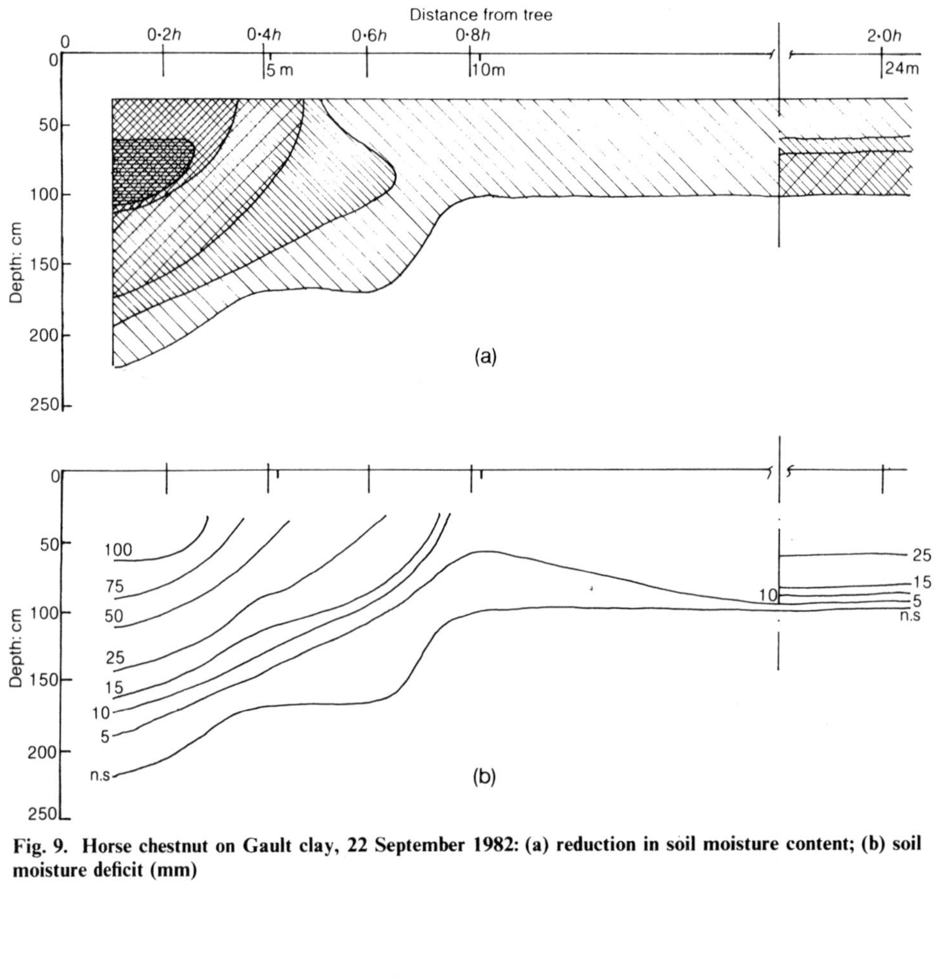

Fig. 9. Horse chestnut on Gault clay, 22 September 1982: (a) reduction in soil moisture content; (b) soil moisture deficit (mm)

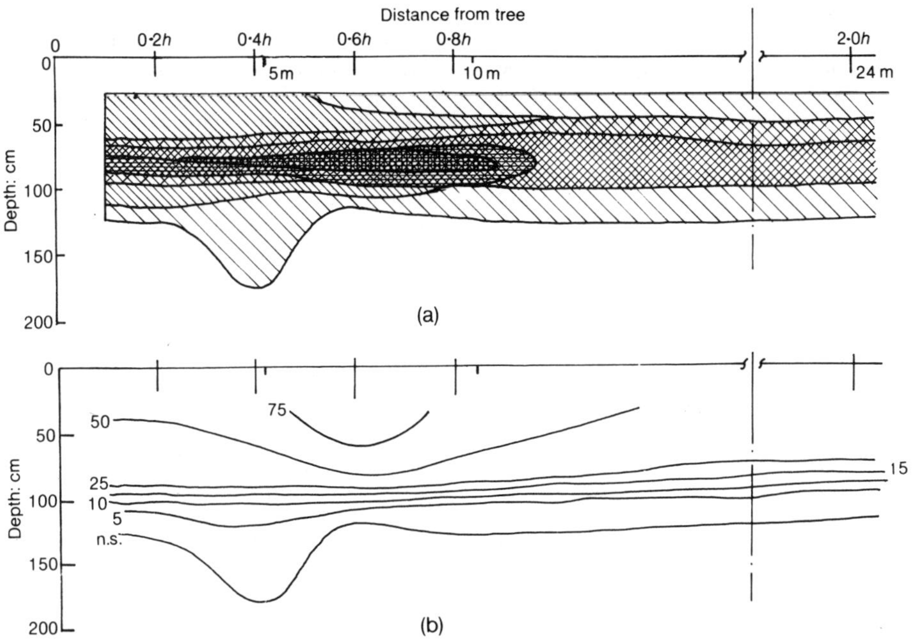

Fig. 10. Silver birch on Gault clay, 22 September 1982: (a) reduction in soil moisture content; (b) soil moisture deficit (mm)

under *Variation between soil types*). Fig. 6 (lime tree on boulder clay) can be compared to Fig. 1 to illustrate this similarity.

The main exception to this is silver birch. Of the four replicates, two are similar and one other has only slight effects which are masked by soil abnormalities. The remaining replicate (on London clay) indicates effects to 300 cm with up to 10% reduction in moisture content between 150 and 250 cm. These effects do not extend beyond 0·4*h*, but suggest exceptional effects by this individual tree. However, there are indications that these unusual effects are being produced by a subsoil drain, rather than root activity.

Variation between soil types

The patterns of deficit produced by the trees do not appear to be influenced significantly by the different clays, with the exception that the differences in clay permeability can influence the depth of moisture deficits with species of very high water demand such as poplar (see under *Influence of soil permeability and permanent water deficits*).

The general similarity on different clays is illustrated by horse chestnut trees in Figs 2 and 7–9. These figures also again demonstrate the similarity between individuals of the same species (see under *Variation between individuals of the same species*).

Influence of abnormalities in soil profiles

A feature of clay soils is their very low rate of lateral or vertical moisture movement. As a result, the water removed by root activity produces a localized deficit in the immediate vicinity. The overall patterns of deficit produced in this way have been illustrated in Figs 1–9.

However, in some examples the uniformity of the clay has been disrupted by bands of sand or other highly permeable material (ballast, gravel etc.) which allow rapid water movements, particularly from adjacent areas. Not only can such water replenish any deficits produced in the clay, but they also provide a more readily available source of water for the tree. As a result, any such band can have a major influence on the patterns of deficit. This is illustrated in Fig. 10 (a silver birch on Gault clay), which has a number of bands of highly permeable sand in the surface 120 cm. As a result, no deficit extends into the underlying clay, and any effects by the tree are masked by the effects of the natural drainage of the bands of sand.

Bands of permeable material of this sort were frequently encountered when inserting the access tubes, particularly in the surface two metres. Approximately one third of the trees on the Oxford and boulder clays in the Milton Keynes area were unsuitable for this reason and were rejected for experimental use. Availability of water in this way might provide an explanation for the occasional anomaly of trees of high water demand in close proximity to buildings which do not cause any foundation movement.

Anomalous results were also produced on sites where there was an adjacent area of permeable soil overlying the London clay. For instance, one of the lime trees on London clay is situated at the base of a gentle (approximately 5°) slope, with Boyn Hill Gravel (a post-glacial deposit) overlying the clay approximately 50 m upslope from the tree. It appears that water discharges from the gravel and runs down on the surface or through the topsoil and results in virtually no deficit developing in the clay, even at the closest positions to this tree.

Influence of soil permeability and permanent water deficits

The results presented in Figs 1–10 show the difference in moisture content between the driest and wettest times during 1982. However, it is also important to consider any moisture deficit which persists from year to year, and which will not be demonstrated by monitoring the seasonal changes.

Ideally any such semi-permanent deficits should be determined by comparison with conditions before the tree was established or after it has died and the ground water returned to field capacity, but there are practical difficulties in such an approach. The alternative is to assume that the normal field capacity close to the tree is similar to that at the comparison hole, and to use the springtime readings at this comparison hole as the basis for contrasting the moisture content profiles near the tree. This assumption produces some discrepancies because of inherent differences in the soil, but it provides the only viable method for determining these long-term effects.

The results fail to show any significant long-term effects for any of the trees, with the exception of all nine of the poplars on the clay soils (the poplar on the low shrinkage clay silt produced anomalous readings caused by bands of sandy gravel).

The difference between the moisture content profiles at the close (0·2*h* or 0·25*h*) and comparison holes is shown for four of the poplars on the different clay types in Figs 11–14. These all demonstrate lower moisture content values close to the tree. For the example on boulder clay (Fig. 11), this extends to 290 cm, with the profiles essentially similar below this level, and on the Oxford clay (Fig. 12) the profiles converge at 350 cm. However, on the London and Gault clays (Figs 13 and 14) there are still significant differences at the bottom of the access tubes at 350 cm, suggesting that the drying effects extend below this level.

Of even greater interest is to contrast the autumn and spring moisture content profiles at the

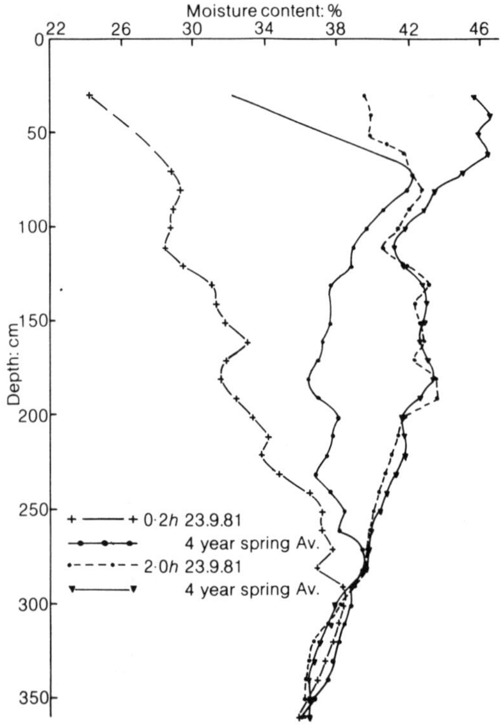

Fig. 11. Poplar on boulder clay: soil moisture content profiles for access holes at 0·2*h* and 2*h* in autumn and spring

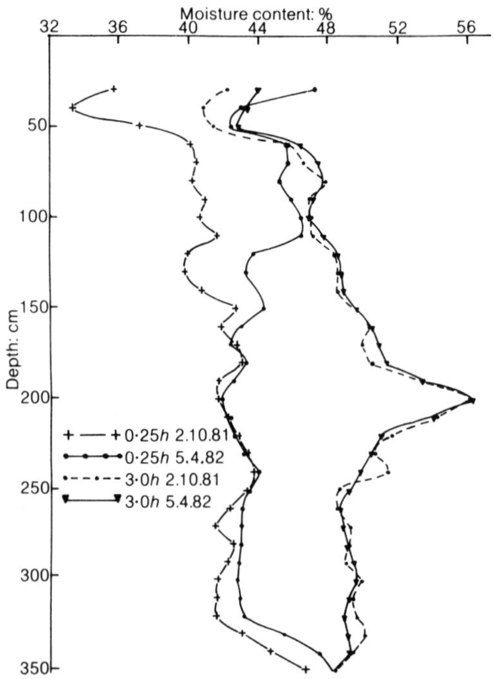

Fig. 12. Poplar on Oxford clay: soil moisture content profiles for access holes at 0·25*h* and 3*h* in autumn and spring

close and distant positions. On the boulder clay (Fig. 11), at the $2h$ distance the profiles converge below 100 cm demonstrating the effects of grass to this depth (see also Fig. 3). Below this level the only differences produced are by the random counting error of the neutron probe. At the $0{\cdot}2h$ distance the spring readings show that moisture has penetrated the full depth of the desiccated soil, that is to 290 cm, although the profile has still not reached the higher moisture contents observed at the $2h$ distance.

This should be contrasted with the Oxford clay (Fig. 12). This also shows the shallow effects produced by grass at $3h$, but at the $0{\cdot}25h$ distance the differences in autumn and spring profiles only extend to 170 cm. This indicates that the winter rainfall is only penetrating to this depth and that no seasonal changes are occurring below this level (except for slight differences in the profiles between 250 cm and 350 cm, which is possibly produced by a rising water table). The profiles indicate a deep-seated deficit which persists from season to season.

Similar effects are seen in London clay (Fig. 13), with the seasonal effects at $0{\cdot}8h$ extending to 160 cm and a semi-permanent deficit below this level. On Gault clay (Fig. 14) the differences appear to extend to 220 cm, but this water penetration is aided by sandy soil to 130 cm at the $0{\cdot}25h$ hole (this accounts for the lateral shift in the moisture content profiles).

These differences between the clay types in the depth of seasonal variation presumably reflect differences in permeability of the clay. The reasonably permeable boulder clay permits rehydration through the full depth of the profile, and a single winter of rainfall is almost sufficient to return the soil to its full field capacity. This moisture is then available for the tree the following year and provides a large reservoir, which may explain why the deficits do not extend below 300 cm. However, on the less permeable clays there is very little replenishment each year, so that further sources of water can only be found by root exploitation to greater depth or radial distance.

These semi-permanent moisture deficits are of great significance to the problems of foundation heave following removal of trees. The surface seasonal variations will occur each year, but if trees are removed the deep-seated desiccated soil will slowly rehydrate and swell, producing the problems of long-term heave. Observations of

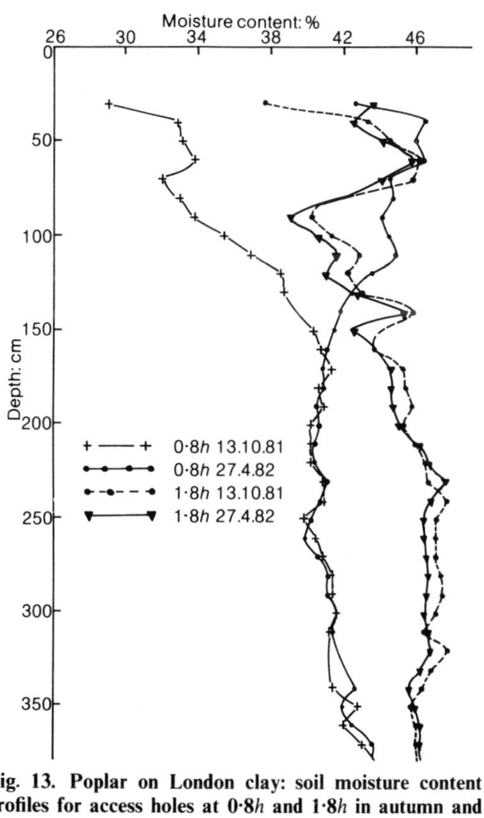

Fig. 13. Poplar on London clay: soil moisture content profiles for access holes at 0·8*h* and 1·8*h* in autumn and spring

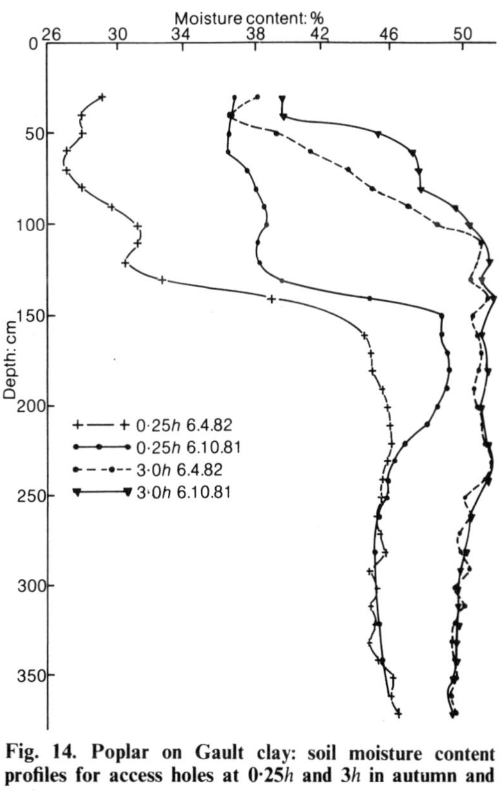

Fig. 14. Poplar on Gault clay: soil moisture content profiles for access holes at 0·25*h* and 3*h* in autumn and spring

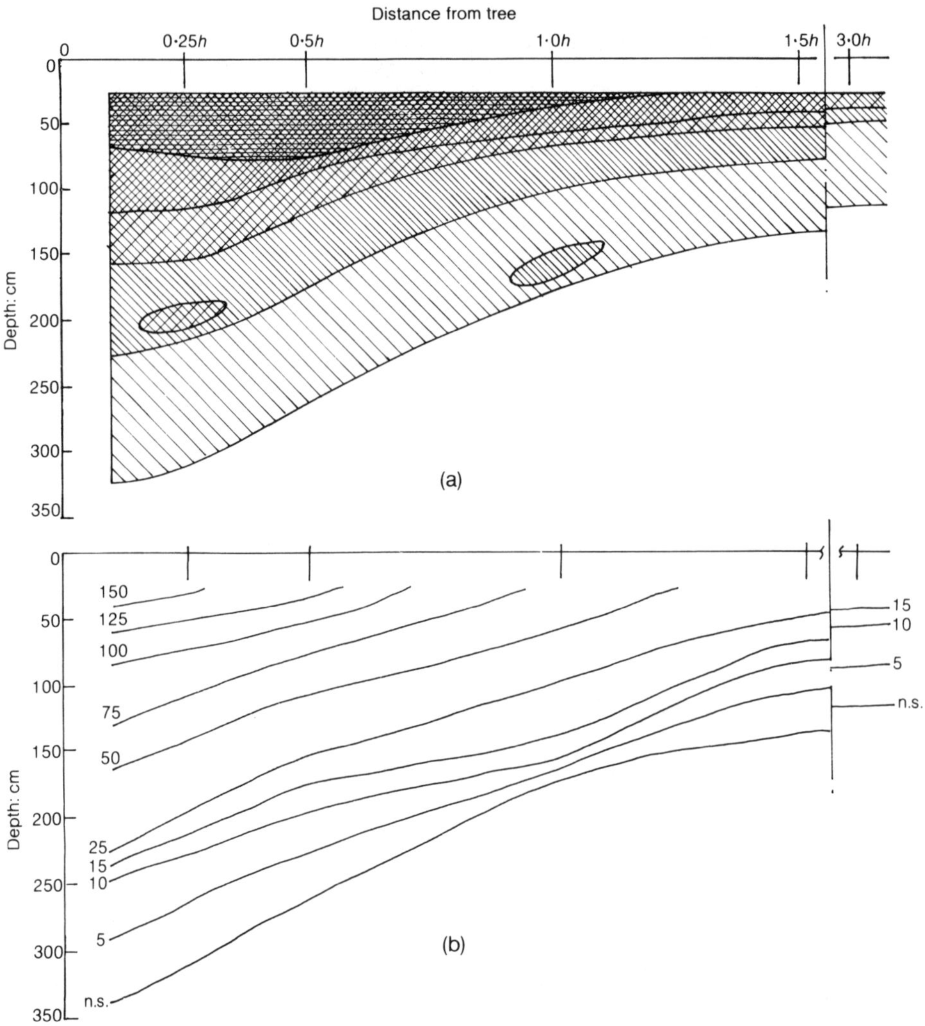

Fig. 15. Poplar, average of nine trees, on boulder/Oxford/Gault/London clays in autumn 1982: (a) reduction in soil moisture content; (b) soil moisture deficit (mm)

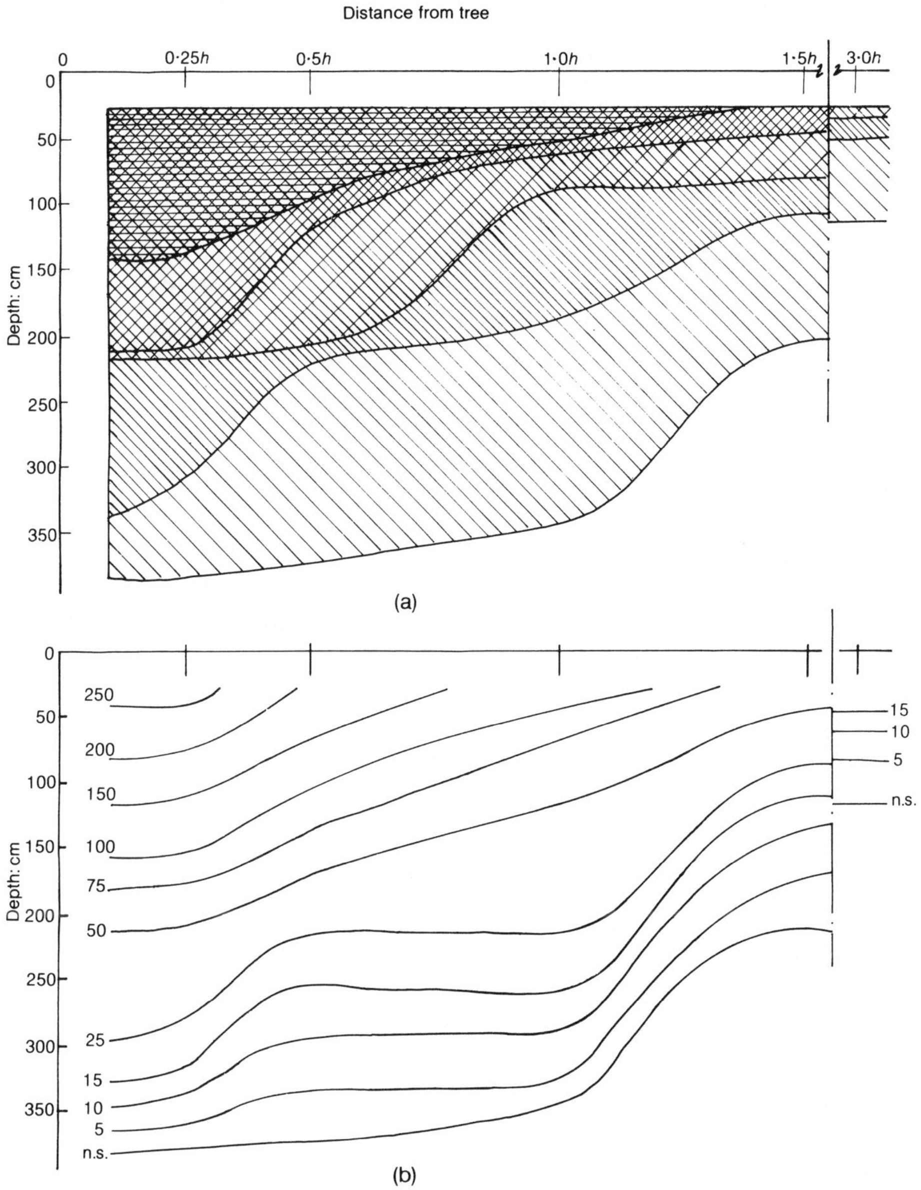

Fig. 16. Poplar, average of nine trees on boulder/Oxford/Gault/London clays in autumn 1982: (a) total reduction in soil moisture content (permanent and seasonal); (b) total moisture deficit (mm)

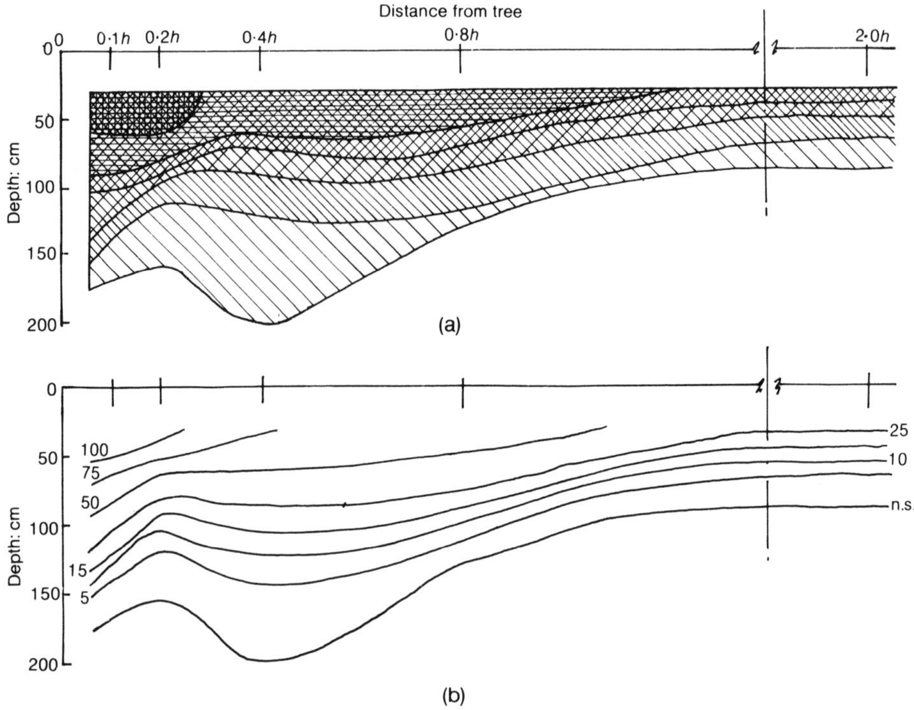

Fig. 17. Lime, average of five trees, on boulder/Oxford/London clays in autumn 1982: (a) reduction in soil moisture content; (b) soil moisture deficit (mm)

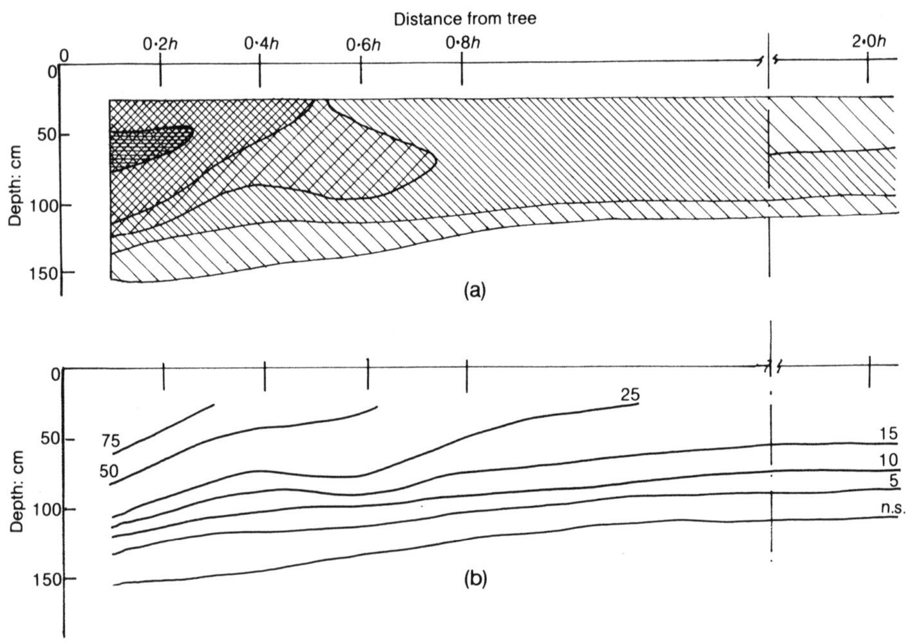

Fig. 18. Horse chestnut, average of ten trees, on boulder/Oxford/Gault/London clays in autumn 1982: (a) reduction in soil moisture content; (b) soil moisture deficit (mm)

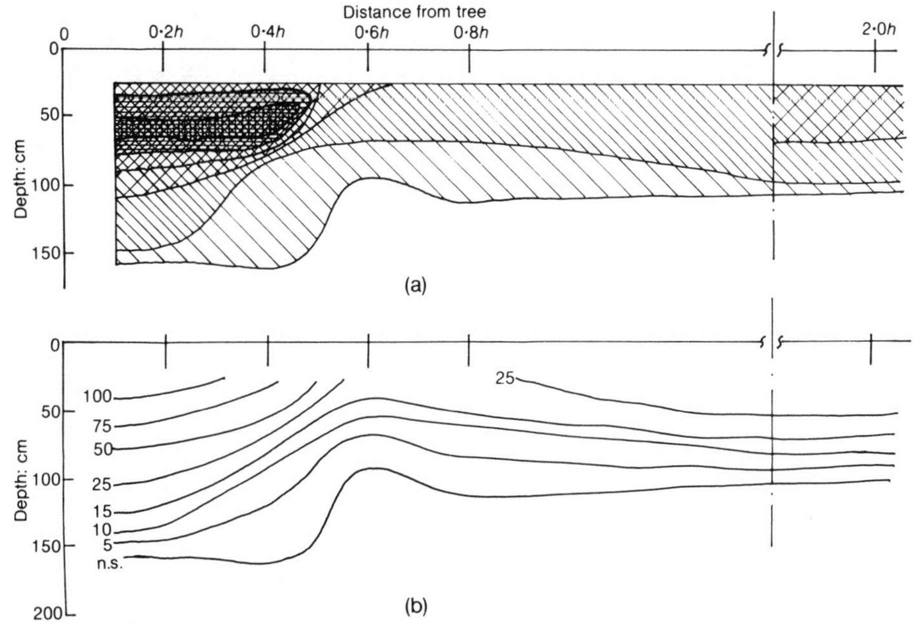

Fig. 19. Silver birch, average of two trees, on boulder/London clays in autumn 1982: (a) reduction in soil moisture content; (b) soil moisture deficit (mm)

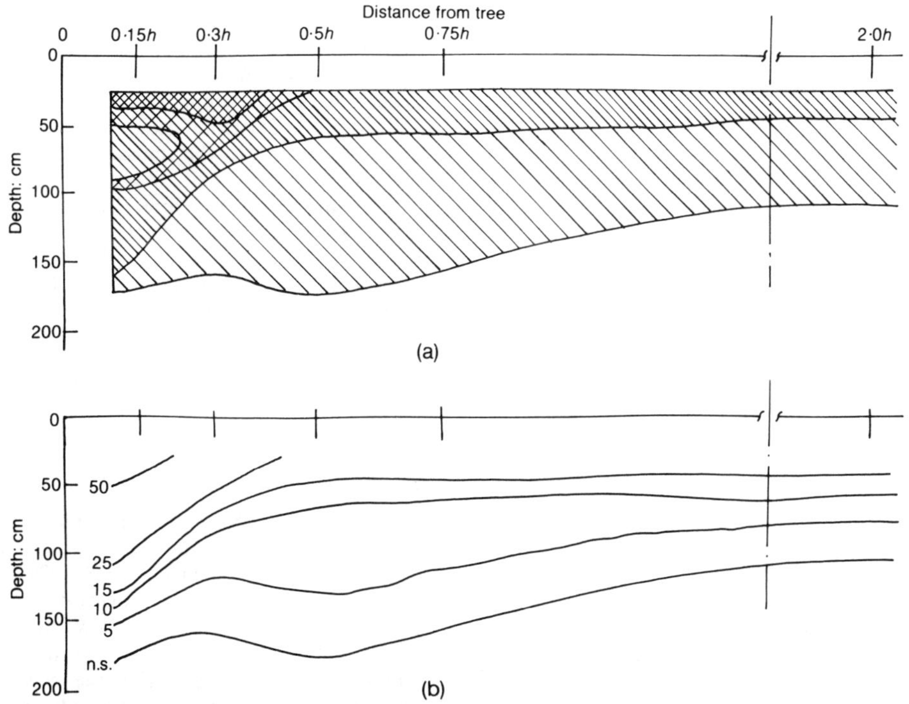

Fig. 20. Leyland cypress, average of six trees, on boulder/Gault/London clays in autumn 1982: (a) reduction in soil moisture content; (b) soil moisture deficit (mm)

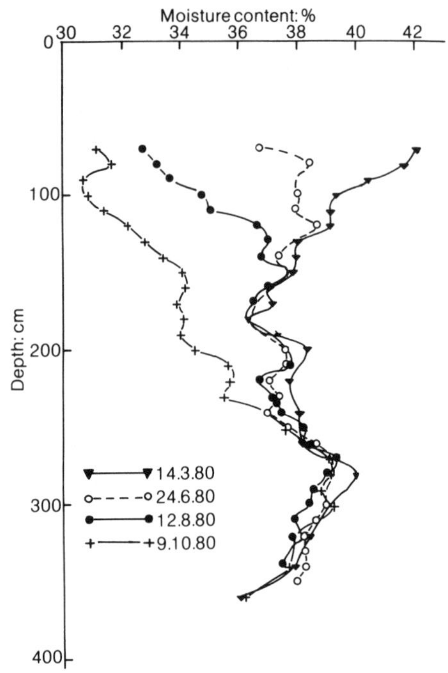

Fig. 21. Seasonal development of soil moisture content profile at 0·2h from poplar on boulder clay

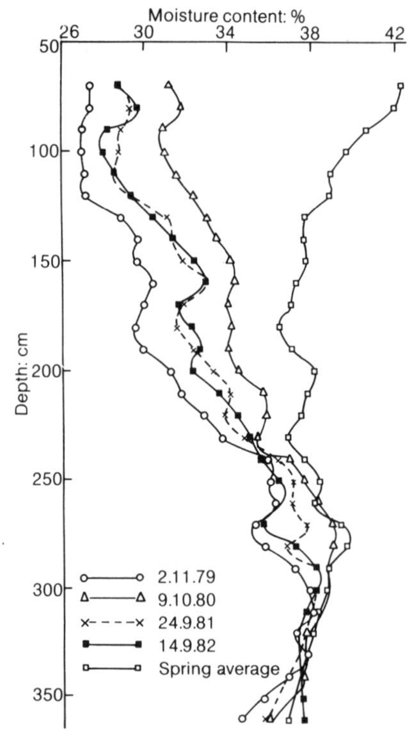

Fig. 22. Poplar on boulder clay: comparison of driest moisture content profiles for different years

a row of cottages on London clay at Windsor showed that soil movements produced by elm trees, which are another species of high water demand, can continue for more than 20 years (Samuels & Cheney, 1975).

It is also important to reiterate that no semi-permanent deficits could be demonstrated for any of the other species in this investigation, which suggests that moderate or low demand tree species do not cause problems of heave. This may not apply in some situations, and a deep-seated deficit may develop closer to the tree than the nearest of our access holes, but it does indicate that the problems of heave occur mainly with species of high water demand on clays of low permeability.

Average results for the different species

Previous sections have noted the general similarity between individuals of the same species, and that there is little difference between soils except for the influence of soil permeability. This suggests that all the results for each species can be grouped together into a single average result. These results are presented in Figs 15–20.

Two average results are shown for the poplars; these are the seasonal effects, and the permanent plus the seasonal effects. Differences in moisture content profiles extend below the base of the access tubes at 380 cm, indicating that permanent soil moisture deficits are even more extreme than those which are shown. Despite this, soil moisture deficits reach a maximum of 250 mm near the tree, and extend to 350 cm at 1h and to 210 cm at 1·5h. This demonstrates the massive effects of this species of tree.

By comparison, the lime and horse chestnut (Figs 17 and 18) show effects only extending to between 150 cm and 200 cm with little or no effect at 0·8h. The silver birch (Fig. 19) is the average of only two trees (two others having to be rejected as a result of the effects of soil abnormalities) and shows a localized and shallow effect. The Leyland cypress (Fig. 20) has a slight and very localized effect.

Development of deficit during the year

Readings have been taken at some of the access tubes at intervals during the year. The seasonal development of the deficit during 1979 at one of the access holes (0·2h from poplar on boulder clay) is illustrated in Fig. 21. This shows the gradually increasing depth and reduction of moisture content which is achieved during the season. Comparison at the different distances from the tree shows that these effects progressively radiate out from the tree during the summer.

Differences of profile between years

Results on some trees are available each year since 1979 and provide an indication of the differences in development of moisture content between years. The profiles at 0·2*h* from a poplar on boulder clay are shown in Fig. 22. This shows that the driest conditions developed during 1979, and the least effect during 1980. In this figure there is little difference between 1980 and 1981, but most of the trees show slightly greater effects during 1981.

These differences between years reflect the summer weather conditions, particularly rainfall. None of the years since the beginning of the research have been particularly dry, and reasonable drought conditions are awaited to demonstrate the increased effects of the trees during more extreme conditions. Once these data become available, efforts will be made to correlate the effects of the trees with meteorological data.

DISCUSSION

These two research projects are providing a mass of information on the effects of different trees and different clay soils. It would be premature to reach conclusions at this stage, particularly before there is a drought of reasonable severity to provide observations on the performance of the trees during such conditions. However, some preliminary conclusions can be drawn.

The most important conclusion is the tremendous difference between poplar and the other species. It has long been known that poplar produces greater effects, and this has been reflected in the NHBC Practice Note 3 (National House-Building Council, 1969) which identifies poplars, elms and willows for more stringent precautions in foundation design. However, the magnitude of these differences had not been demonstrated previously.

This is of particular significance when considering the evolution of existing guidelines on foundation design near trees. In the investigations by Ward (1953) which gave rise to the concept of separating trees and buildings to distances equal to their height, 'most of the trees are poplars and elms, oak is fairly frequent and others are alder, willow, ash, sycamore, chestnut, birch and hawthorn'. Experience indicates that elm and probably oak are similar to poplar and that these three species (possibly plus willow) should be identified as high water demand species. The guidelines which are still in use today are based primarily on investigations on these three species, but despite this are applied to all species of tree, irrespective of their water demand.

New guidelines are required which correctly identify the greater effects of high water demand species, and also make provision for the lesser effects of species of particularly low demand (of which birch, beech and holly are probably examples).

Although the results have not demonstrated any significant differences in development of moisture deficits on different soils (except for the influence of permeability on deep-seated deficits), differences in clay type are still important. This is because the amount of soil movement which is produced by the moisture deficit will depend on the clay shrinkage characteristics. These differences in clay type are recognized in Digest 240 (Building Research Establishment, 1980a) which recommends that shrinkage potential should be defined according to the plasticity index of the clay. Thus, although patterns of soil moisture deficit may be similar for different clays, the amount of clay shrinkage and the associated risk of structural damage will depend on the clay characteristics. Work is in hand to measure the amounts of soil movement which are occurring adjacent to some of the access holes, and to correlate these with the plasticity index of the clay.

The introduction to this Paper drew attention to the change in emphasis in recent years from the concept of separating trees and buildings to that of accepting that trees will be planted or grow near buildings and designing the foundations accordingly. At present this concept is unacceptable to the building trade, because of the fear that it will impose excessive extra foundation depths, similar to those of the existing NHBC Practice Note 3 (National House-Building Council, 1969).

The research described in this Paper may help to allay such fears. Extreme depths are required for high water demand species on very high shrinkage clays, but such a combination of conditions is exceptional. For most tree species (except poplar, elm, oak and willow), a modest increase in foundation depth will permit trees to be planted as close to the building as desired (other factors, such as loss of light, leaves falling into gutters or direct obstruction of the branches also influence the desirable proximity of tree growth). The present results, albeit without the additional information from a drought year, combined with other practical experience, suggests that an increase in foundation depth to 1·5 m on very high shrinkage clay should be sufficient to accommodate most tree planting designs with species of moderate water demand. This depth could be reduced for clays of lower shrinkage potential, with the existing standard 1 m deep foundation being adequate for the nominal effects of low shrinkage clays.

The important concept is that the soil type should be assessed prior to design and building, and the appropriate extra foundation depth incorporated at that stage. Thereafter the householder, as well as his insurers, surveyors etc., can rest assured that reasonable tree growth can be tolerated without undue risk of structural damage and without constant fear and pressure for removal of all trees within their height of the building. Particular precautions need only be taken with high water demand species. New guidelines of this sort, which incorporate allowance for soil type and water demand of the tree, need to be adopted and then enforced at planning and building control stages.

In some situations this will entail some extra cost on foundations works, but this need not be as great as builders currently fear, and this would be more than compensated by the improved performance of the building and aesthetic benefits to the urban treescape in shrinkable clay areas.

ACKNOWLEDGEMENTS

The research described in this Paper is funded by the National House-Building Council and by Milton Keynes Development Corporation, whose support is gratefully acknowledged. The views expressed are personal, and do not necessarily represent the views of either sponsor.

REFERENCES

Bell, J. P. (1976). *Neutron probe practice*. Report 19. Wallingford: Institute of Hydrology.

British Standards Institution (1980). *Code of practice for trees in relation to construction*. BS 5837. London: British Standards Institution.

Building Research Establishment (1980a). *Low rise buildings on shrinkable clay soils: part 1*. Digest 240. London: HMSO.

Building Research Establishment (1980b). *Low rise buildings on shrinkable clay soils: part 2*. Digest 241. London: HMSO.

Building Research Station (1949). *Foundations on shrinkable clays*. Digest 3. London: HMSO.

Cutler, D. F. & Richardson, I. B. K. (1981). *Tree roots and buildings*. Harlow: Construction Press.

National House-Building Council (1969). *Root damage by trees—siting of dwellings and special precautions*. Practice Note 3.

Reynolds, E. R. C. (1979). *Tree roots and built developments—a review*. Report to the Department of Environment. Contract F3/2/148.

Samuels, W. G. & Cheney, J. E. (1975). Long-term heave of a building on clay due to tree removal. *Proc. Conf. Settlement of Structures, Cambridge*, 212–220. BGS Symposium, 1974, London: Pentech Press.

Ward, W. H. (1953). Soil movement and weather. *Proc. 3rd Int. Conf. Soil Mech., Zurich* **4**, 477–482.

The effects of vegetation on the swelling and shrinking of soils in Australia

B. G. RICHARDS,* P. PETER* and W. W. EMERSON*

Soil movement is of great importance in Australia because about 20% of the total area consists of expansive soils, most of which occur in the semi-arid climatic zone. The first measurements of soil movement were made in Adelaide under natural grassland. From this limited information, it was deduced that there was little soil movement below 1·6 m. Recently, movements have been measured in clay layers at much greater depths. These movements have been attributed to shrinkage due to desiccation by deep rooted vegetation, and swelling due to ponding of water. Several case histories are discussed where the effects of vegetation have caused movements of buildings of up to 150 mm settlement and 100 mm heave. To calculate potential soil movement caused by vegetation, the extremes of total soil suction and a quantitative value of the volume change properties of the soil are required. The instability index, $I_{p''}$ has been suggested as a measure of this soil property and was measured on two of the most expansive and widespread soil types in Adelaide. Values for the black earths were 0·06 and for the underlying dispersive clay, 0·10, compared with 0·02 for the red-brown earths down to 5 m. Changes in total soil suction were measured in these two soils both near and away from several species of trees. Near eucalyptus trees, the total suction of red-brown earths attained a maximum value of 2 MPa to depths exceeding 5 m. Under natural grassland, the profile was uniformly wetted to a suction of 0·3 MPa. In the saline dispersing clay underlying the black earths, the corresponding suctions were of the order of 3 MPa and 1 MPa respectively. However, in this latter case, the effect of $I_{p''}$ and the depth of the clay, causes much greater movements in spite of the lower suction changes. At present, soil movements due to the effects of vegetation and its associated watering regime, calculated from measurements of $I_{p''}$ are only a guide. Past experience of the movements of buildings on similar soil profiles in similar circumstances is still essential.

Le mouvement du sol est très important en Australie, parce que 20% environ du continent entier comporte des sols expansibles, dont la plupart se trouvent dans la zone climatique demi-aride. Les premières mesures du mouvement des sols furent prises à Adelaide sous des herbages naturels. Sur la base de ces données limitées on a déduit qu'il n'y avait que peu de mouvement du sol à une profondeur supérieure à 1·6 m. Récemment des mouvements ont été mesurés dans des couches d'argile à des profondeurs beaucoup plus grandes. Ces mouvements ont été attribués au retrait causé par la dessiccation due à la végétation profondément enracinée, tandis qu'une autre cause possible est le gonflement produit par la formation de mares. L'article discute plusieurs exemples où les effets de la végétation ont causé des tassements jusqu'à 150 mm et des soulèvements jusqu'à 100 mm de bâtiments. Pour calculer le mouvement potentiel du sol produit par la végétation, il faut connaître les valeurs extrêmes de la succion totale du sol et une valeur quantitative de ses propriétés de changement de volume. L'indice d'instabilité $I_{p''}$ a été proposé comme mesure de cette propriété des sols, et cette valeur a été déterminée sur deux des sols les plus expansibles et les plus usuels à Adelaide. Les valeurs pour les terres noires ont été de 0·06, la valeur pour l'argile sousjacente dispersive étant de 0·10, à comparer avec 0·02 dans le cas des terres brunes-rougeâtres jusqu'à une profondeur de 5 m. Des changements dans la succion totale du sol ont été mesurés dans ces deux sols dans le voisinage d'arbres de plusieurs essences, ainsi qu'à quelque distance. Dans le voisinage des eucalyptus on a trouvé une valeur maximale de 4 MPa pour la succion totale des terres brunes-rougeâtres, jusqu'à des profondeurs dépassant 7 m. Sous des herbages naturels le profil a été humidifié de façon uniforme à une succion de 0·3 MPa. Dans l'argile dispersive saline sous les terres noires les succions correspondantes étaient repectivement de l'ordre de 1 MPa et 3 MPa. Cependant, dans le dernier cas l'effet de $I_{p''}$ et la profondeur de l'argile causent des mouvements beaucoup plus marqués, malgré les changements réduits dans la succion. A present les mouvements du sol dus aux effets de la végétation et au régime d'eau associé, calculé à partir des mesures du $I_{p''}$ ne donnent qu'une indication générale. On a encore essentiellement besoin de données concernant les mouvements des bâtiments dans le passé sur des profils de sol similaire dans des circonstances analogues.

*Commonwealth Scientific Industrial Research Organization, Division of Soils, Glen Osmond, South Australia.

NOTATION

- h soil suction
- $I_{p''}$ instability index
- m total number of soil layers
- n variable representing soil layer
- p_0 overburden pressure
- z depth below soil surface
- ε strain

INTRODUCTION

The problems associated with engineering structures built on swelling clay soils are very important, but are not adequately understood. This fact has been demonstrated in four international conferences on expansive clays held since 1965. The problems in Australia are particularly significant for the following reasons: the extent of swelling clay soils, the seasonal aridity of the climate where most of these soils occur and the severity of swelling.

The last reason is due to the high surface area of the clay fraction present in such soils coupled frequently with the presence of exchangeable sodium and magnesium ions.

GEOGRAPHIC DISTRIBUTION OF SWELLING CLAY SOILS

Australia is a very old continent geologically and erosion has created large areas of alluvial landscapes. Where drainage is limited, the soil profiles on these landscapes are often clays to the surface and very deep, such as the black earths (Stace, Hubble, Brewer, Northcote, Sleeman, Mulcahy & Hallsworth, 1968). When these profiles are derived from basalt, the clay fraction contains predominantly montmorillonite (Norrish & Pickering, 1977). Other less expansive clay soils which occur on the gentle slopes, such as red-brown earths (Stace *et al.*, 1968), also contain fine grained illite and interstratified illite/montmorillonite, but the total amount of clay in the profile and the surface area of the clay is smaller. Black and red-brown earths are included together with those other great soil groups considered at least moderately expansive in Fig. 1 to show the distribution of moderately to highly expansive clay soils in Australia. Fig. 1 is derived from the soil map of Northcote, Hubble, Isbell, Thompson & Bettenay (1975), which is based on the conventional soil profile used for classification by pedologists, that is to a depth of about 1 m

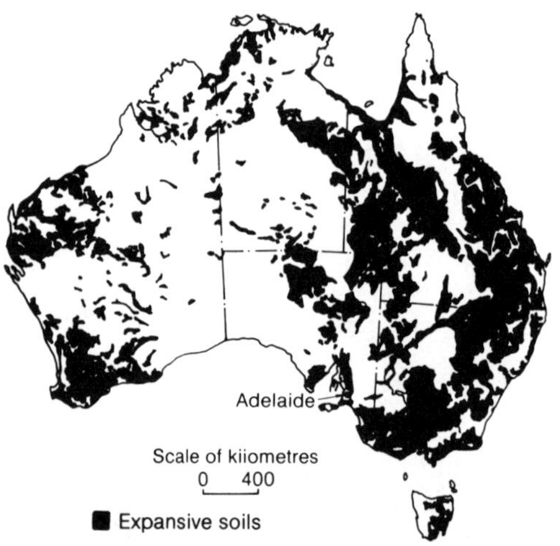

Fig. 1. Distribution of moderately to highly expansive soils in Australia

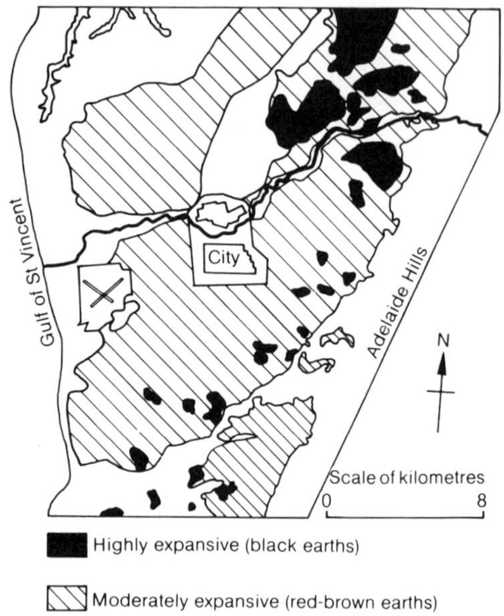

Fig. 2. Expansive soils of the Adelaide region

(Northcote, 1974). The expansive soils cover about 20% of the total continent. In Fig. 2, a map of the urban area of Adelaide is presented showing the large areas of red-brown earths (slightly to moderately expansive) and black earths (highly expansive) which underlie that city (Selby, 1979).

This Paper is concerned principally with experience of how vegetation causes movement in these two soil groups. At the same time, the importance of water content changes in the soil layers underlying the conventional pedological soil profile will be emphasized.

SWELLING CLAY SOILS AND CLIMATE

Aitchison & Richards (1965) have used a simplified Thornthwaite climatic index (Russam & Coleman, 1961), to delineate the climatic zones of Australia (Fig. 3). The index uses an estimate of the drainage or moisture deficit of a hypothetical soil profile using monthly rainfall and potential evaporation data. High positive values indicate net drainage with no significant desiccation (i.e. wet or humid zones) whereas high negative values, the reverse (i.e. arid zones). Aitchison & Richards (1965) found a rough correlation between the upper limit of the equilibrium total soil suction under covered areas and this index.

The shaded area in Fig. 3 indicates an arbitrary definition of the semi-arid zone of Australia namely where the index lies between +20 and −40. This zone contains most of the expansive clays of Australia (compare Figs 1 and 3). Similar problems due to expansive clays occur in similar climatic zones in other countries such as the USA and South Africa (Aitchison, 1965).

VEGETATION AND TOTAL SUCTION IN A SEMI-ARID CLIMATE

In the semi-arid zone, transpiration by vegetation will cause desiccation of the soil profile in addition to that due to direct loss of water vapour from the soil. The latter is important in expansive soils with clay to the surface, as deep cracks often in excess of 2 m depth and open to the surface are formed on drying. The severity and depth of drying by vegetation will depend on the type of vegetation. The classical measurement for agricultural crops is the water content of the soil at which sunflowers permanently wilt. This water content has been shown for many soils to be equal to the water content of disturbed samples in equilibrium with a matrix suction of 1·5 MPa (Richards & Weaver, 1943). However, in the arid areas of Australia, some plants have been shown to remain healthy even though the total suction in the root zone approaches 31 MPa, for example Mulga scrublands and spinifex plains (Aitchison & Richards, 1965). The maximum total suction achieved under grass and trees in a semi-arid climate has been obtained using the thermistor psychrometer (Richards, 1965; Peter & Martin, 1973).

At the wet end of the climatic scale, the reference water content is often taken to be that of soil a day or two after it has been thoroughly wetted and allowed to drain. This water content is referred to as the field moisture condition or field capacity. The water content retained by an undisturbed soil sample at a matrix suction of 0·10 MPa is an approximation to field capacity. Therefore, 0·10 MPa is often assumed to be the minimum matrix suction of a soil profile except where a water table occurs at a depth much less than 3 m. The minimum total suction corresponding to field capacity is the sum of 0·10 MPa and the solute suction corresponding to the soluble salts present. For a non-sodic red-brown earth in Adelaide, Emerson & Peter (1981) reported a minimum value of the total suction of 0·2 MPa. From the experimental data now being obtained from the black earths in Adelaide, the corresponding value appears to be 0·3 MPa.

EFFECT OF IRRIGATION IN A SEMI-ARID CLIMATE

To improve the environment in semi-arid zones, lawns and ornamental trees are frequently planted in gardens around houses and public buildings. To maintain these gardens during the dry season, irrigation must be used extensively. This may cause additional swelling of expansive soils compared with natural rainfall conditions for two reasons. Firstly, watering is generally not closely controlled, usually tending to the use of excessive water, so that with time, the whole profile becomes wetted up. Secondly, the concentration of soluble salts is reduced. This will cause those clay soils, class 1, which disperse in water (Standards Association of Australia, 1980) to take up additional water as compared with class 3 soils. Such dispersive clay layers occur widely at depth over Australia, including the Adelaide area, e.g. Keswick clay (Stapledon, 1970).

HISTORY OF RESEARCH IN ADELAIDE

1948–55

Research into swelling soils really began in Australia in the 1940s in Adelaide, South Australia. The location was due to a number of factors. Firstly, the majority of the then urban area was located on soils which were classified as predominantly swelling soils (see Fig. 2). In addition to the soils problems, most houses were built using shallow strip footings. Further, South Australia had a lack of suitable structural timber, resulting in a predominance of solid masonry homes. Consequently, by the 1940s a severe problem had arisen with the cracking of buildings on the swelling soils. The Commonwealth Scientific Industrial Research Organization (CSIRO) Division of Soils, located in Adelaide was then asked to investigate the soil physics and

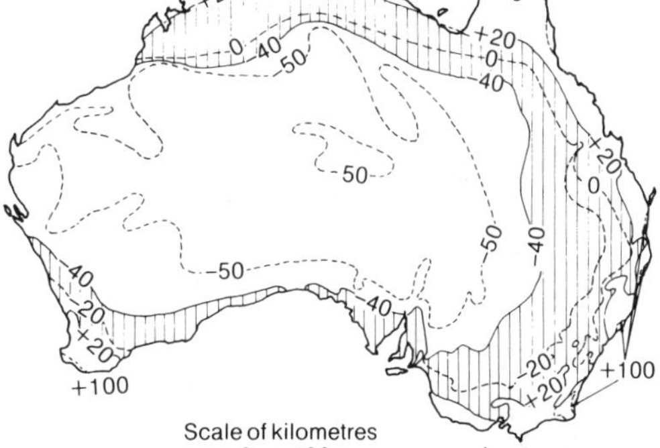

Fig. 3. Climatic map of Australia based on the Thornthwaite index. *I* which is equal to $(100D - 60d)/Ep$, where D = annual drainage and d = moisture deficit of a hypothetical soil profile with a water storage of 4 in using monthly rainfall and potential evaporation data. *Ep* is the annual potential evaporation.

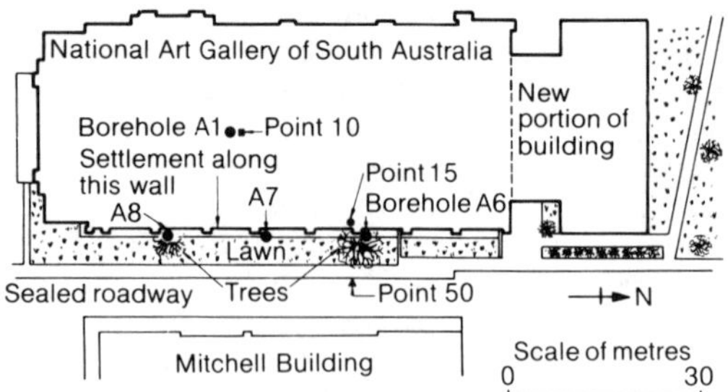

Fig. 4(a). Feature plan of the National Art Gallery building and surroundings

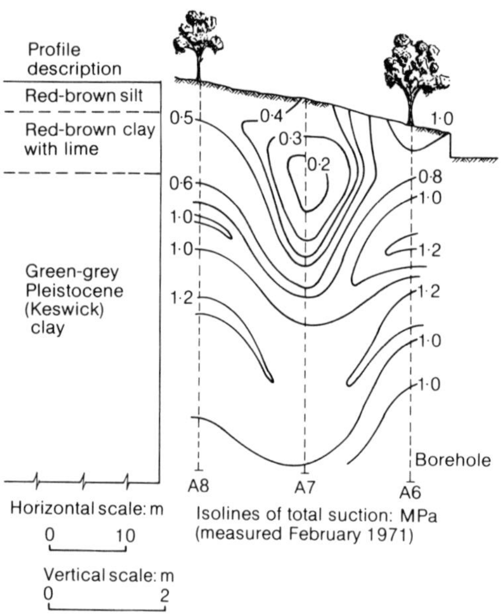

Fig. 4(b). Soil profile and contours of total suction adjacent to the eastern wall of the Gallery before the trees were removed

engineering aspects of the problem. Aitchison & Holmes (1953a, 1953b) concentrated on measurements in open grassed areas of the movements in soil profiles due to seasonal changes in water content. They also selected what they considered to be the most expansive of the soil types within each great soil group, namely the BE1 of the black earth and the RB3 of the red-brown earth (Aitchison, Sprigg & Cochrane, 1954). They measured seasonal ground movements relative to a bench mark at 1·8 m of 79 mm in the BE1 and 35 mm in the RB3. Movements in the other soil types were much less than 10 mm.

The depth of the bench mark of 1·8 m was probably chosen as seasonal moisture changes under grassed areas in England and Australia had been shown to be less than this depth (Ward, 1953; Aitchison & Holmes, 1953a). Because the other factors mentioned before were not considered, they concluded that moisture changes and therefore ground movements were confined to the upper metre or so. This in turn led to the decision to correlate foundation performance with pedological soil types (Aitchison *et al.*, 1954) where depths of the order of only 1–2 m were considered.

1955–65

The early research had a major impact on foundation design in Australia during the next 30 years. During the 1950s Adelaide's water supply improved in quantity if not in quality and corresponded with a change to vegetation which required frequent watering, such as soft grass lawns. This led to some problems with edge uplift due to localized garden watering. For example, Schumann (1967) shows that the total soil suction was maintained near the edge of a house on strip footings with ventilated timber floors at less than 0·3 MPa all year. Under the house, the equilibrium suction remained at 4 MPa. Differential movements of up to 70 mm were measured. Willoughby

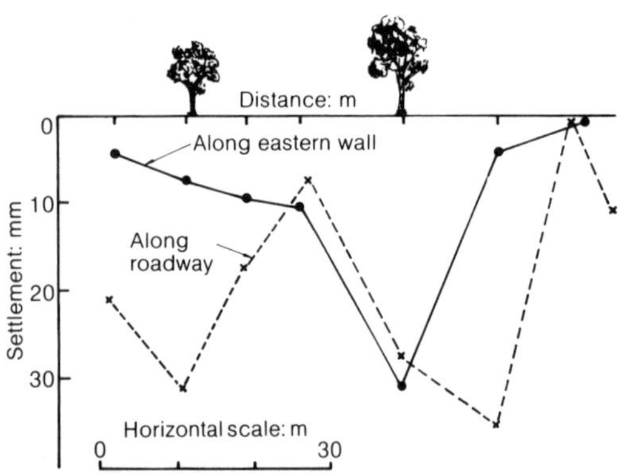

Fig. 4(c). Pattern of settlement along the eastern wall of the Gallery and adjacent roadway before the trees were removed (23 October 1968–10 May 1971)

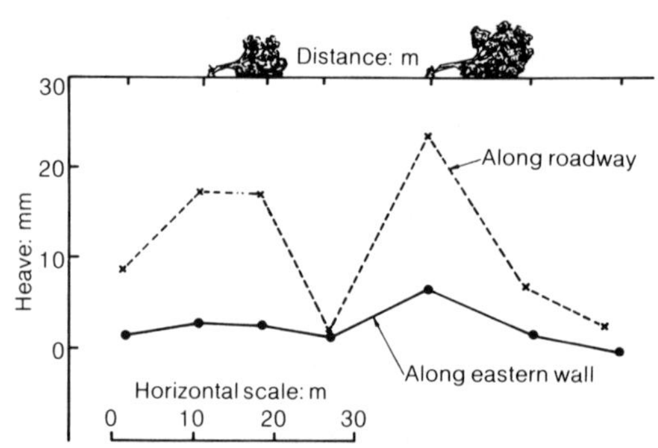

Fig. 4(d). Pattern of heave along the eastern wall of the Gallery and adjacent roadway after the removal of the trees (27 August 1971–8 November 1972)

(1965) also showed that hydrological changes caused by urbanization with its increased garden water, removal of trees, increased runoff from paved and covered areas caused decreases of soil suction and differential heaves.

1965–present

By the late 1960s, gardening trends began changing again and growing the larger Australian native species of vegetation became fashionable. This was partly due to the rising cost of reticulated water and the knowledge that Australian bush gardens can tolerate greater soil suctions (although they actually use more water if it is available). At that time, a number of public buildings in Adelaide, e.g. schools, courthouses and hospitals, began to show signs of serious cracking due to soil movements. Some of the buildings were old ones, built as early as the nineteenth century, and had shown no previous evidence of cracking. CSIRO was asked to assist in the investigation and monitoring of the behaviour of some of these buildings, due to its previous experience in the Adelaide area. A summary of these investigations was published (Aitchison, Peter & Power, 1977), and results obtained at three sites where vegetation was a major factor are discussed.

CASE HISTORIES

One of the most interesting cases is the cracking of the National Art Gallery of South Australia. The gallery was built during the late 1890s and for the next 60 or more years showed no evidence of significant cracking. During a reconstruction of the surrounds of the gallery in the early 1960s, trees were planted along the eastern wall. The plan view of the gallery and the location of the trees are shown in Fig. 4(a). The cracking of the gallery began to cause concern early in 1969, when the investigation commenced. The soil profile and the contours of total soil suction at that time are shown in Fig. 4(b). The soil suction profiles both near and away from the trees are shown in Fig. 5. Levels were also taken at regular intervals on the building and the adjacent roadway, as shown in the time–movement plots in Fig. 6. At the end of 1971, a decision was made to remove the trees and the effect of this removal is clearly shown in Fig. 6. The pattern of movements along the eastern wall of the gallery and the adjacent roadway before and after the removal of the trees are also shown in Figs 4(c) and 4(d). This study clearly shows the role of trees in this case, and how a partial recovery of the differential settlements was achieved by the removal of the trees.

The Barwell flats (a series of 3 storey buildings) and the Modbury High School (a 2 storey complex) were both built during the 1960s on piers 3·65 m and 1·8 m deep respectively, located below black earth profiles in Pleistocene clay (dispersion class 1) which extends to depths several metres below the base of the piers. Instead of a small settlement as anticipated both structures were initially damaged by cracking resulting from the differential heave. As a result of watering newly established gardens and changing hydrology due to urbanization of the areas, heaves of the order of 60 mm and 100 mm were reported by Aitchison *et al.* (1977) for the flats and school respectively. In both cases, some ten years later, trees and large shrubs began to dominate the boundary conditions with the result that the initial heave was reversed near where they were planted. At Modbury High School, a subsequent differential settlement of up to 50 mm has been measured. This is now causing additional damage to the building.

A movement installation near Modbury High School is showing a similar pattern. An initial maximum surface heave of over 30 mm from 1969 to 1975 and a final settlement from 1975 to 1982 of over 40 mm were measured relative to a bench mark at depth of 12 m. The corresponding movements on the deepest rod at 3 m depth was an initial heave of 28 mm and a final heave of 13 mm (i.e. a settlement of 15 mm from the time of the maximum heave in about 1975). This clearly shows that both the changing hydrological conditions

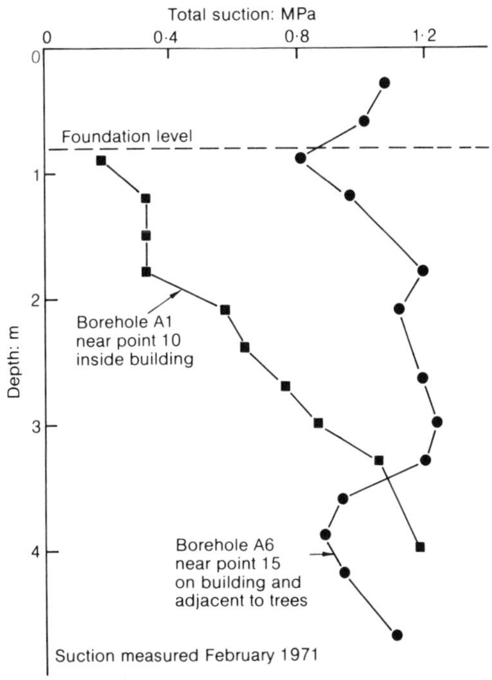

Fig. 5. A comparison of total suction profiles near and away from the trees at the National Art Gallery building

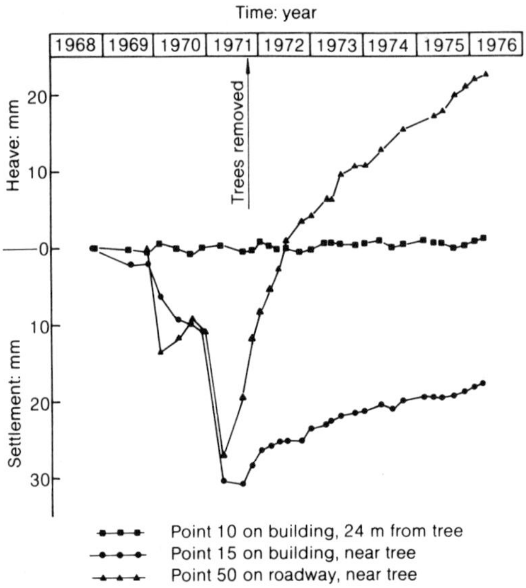

Fig. 6. Time–movement plots of several points inside and one point outside the National Art Gallery building

and the subsequent effects of the vegetation significantly affected the Pleistocene clay below the 3 m rod.

DEFINITION OF PROBLEM

The main problem associated with vegetation on expansive clays is the desiccation it causes in the clay resulting in shrinkage and differential settlement. The settlement or heave of a soil profile due to any change in soil suction is

$$\delta = \sum_{n=0}^{m} \frac{\partial \varepsilon}{\partial \log h} \Delta \log h \, \Delta z_n \tag{1}$$

where δ is the vertical settlement or heave, ε is the vertical strain in soil, $\Delta \log h$ is the change in soil suction ($\log h$ is used to make the variables more linear), Δz_n is the thickness of the nth soil layer and m is the total number of layers.

Following the terminology of Aitchison (1973), equation (1) can be written as

$$\delta = \sum_{n=0}^{m} I_{p''} \Delta \log h \, \Delta z_n \tag{2}$$

where $I_{p''} = \partial\varepsilon/\partial \log h$. $I_{p''}$ has been defined as the instability index, which is analogous to the compression index C_c for consolidation processes and is an intrinsic property of the soil. A similar instability index can also be defined in terms of volumetric strain. It is not a constant, but is a function of many factors, including stress state, soil matrix and solute suction, previous stress and moisture history. However, like other soil mechanics properties an approximate value can be determined in swell–shrinkage tests to represent the in situ soil conditions appropriate to the problem. Obviously for vegetation to cause practical problems, $I_{p''}$ must be significant, e.g. a value of 0·01 may be significant depending on depth of clay and the soil suction change.

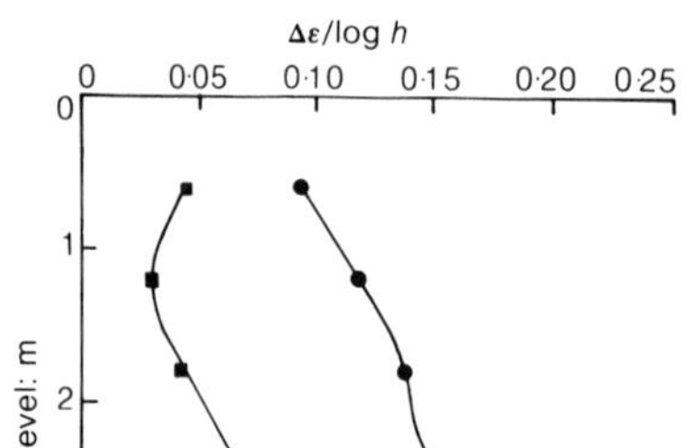

Fig. 7(a). Linear instability index ($I_{p''}$) and volumetric instability index for a typical black earth profile by using the pressure membrane method

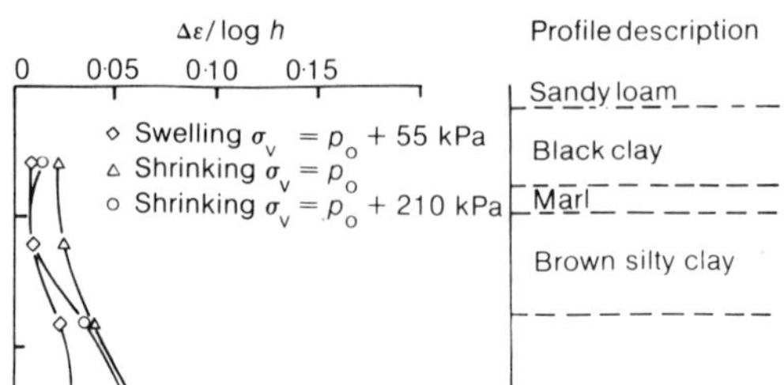

Fig. 7(b). Linear instability indices as determined for shrinking and swelling conditions for a typical black earth profile using a suction controlled consolidometer method (P_0 = overburden pressure)

The other factor in the prediction of ground movement using equation (2) is the prediction of the changes of soil suction, that is $\Delta \log h$. Extensive investigations have been carried out into the redistribution of soil moisture in expansive clays and other soils beneath impervious covered areas throughout Australia and has been described in detail elsewhere (Aitchison & Richards, 1965; Richards, 1967). Of all the mechanisms causing redistribution of soil moisture, one of the most serious and most common of those observed in Australia, including the case histories mentioned, is that due to localized desiccation by large trees adjacent to structures.

CURRENT RESEARCH

Soil characteristics

The problem with expansive clays can be divided into two aspects, namely the expansive characteristics of the soil and the change in soil suction due to the environmental and man-made factors such as vegetation. The early work in the Adelaide area undertaken by CSIRO, Division of Soils and the Department of Mines and Energy of South Australia concentrated on the first aspect, that is the study of the characteristics of the soils of Adelaide (Aitchison *et al.*, 1954; Taylor, 1974).

This early work used the pedological profile system to identify and classify the soil profile, which could then be correlated with foundation requirements using experience. This approach to foundation design was used in Adelaide for nearly 30 years, but recent studies have shown that the pedological system, which was based on sampling to a depth of 0·6–2·0 m is not adequate. Case histories such as those reported have clearly shown that factors such as vegetation, hydrological changes and broken water pipes can influence moisture conditions below the depth of seasonal moisture variation, on which the pedological system is based. As a result of this knowledge, a new study has begun on the soil profiles of the Adelaide urban area to depths of 10 m or to rock, whichever is the least. The aim of this study is to develop a new profile classification system for geomechanics purposes.

The preliminary results of this new study have proved to be interesting. For example, the black earth and red-brown earth profiles of the pedological system, which was stated to be potentially the worst for foundation movement, were re-examined. A black earth profile at the Athelstone Park Primary School (type BE1) was selected as representative of those in an area where some of the largest movements in Adelaide have been observed. The soil profile together with some measured instability indices is shown in Figs 7(a) and 7(b).

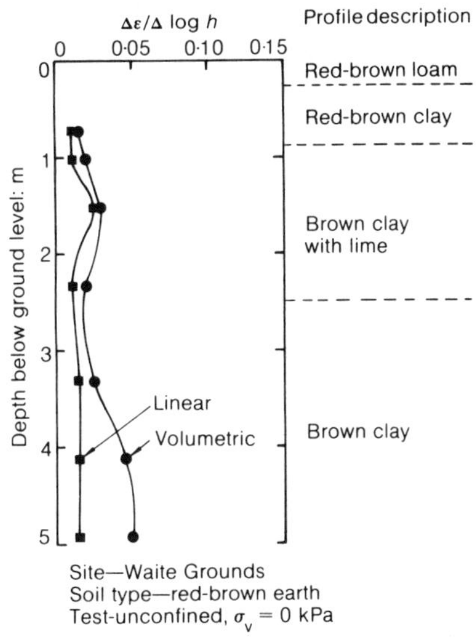

Fig. 8. Linear instability index ($I_{p''}$) and volumetric instability index for a typical red-brown earth profile by using the pressure membrane method

Instability indices $I_{p''}$ were measured by two separate methods on all layers of the profile. The first method, also designed as a practical method, involved placing replicate samples at their in situ moisture condition in pressure membrane apparatus at matrix suctions of 0·03, 0·10, 0·31, 0·98 and 3·10 MPa, with the solution kept approximately constant. The average slope of the volumetric and vertical strain against log suction is used to calculate values of $I_{p''}$, as shown in Fig. 7(a).

The second method used suction controlled consolidometers (Aitchison, Peter & Martin, 1973), in which the sample was taken through various stress and suction paths. This method takes up to 2 years per sample and must be considered a research method. The results using this method are shown in Fig. 7(b). In the laboratory determinations of $I_{p''}$, changes of total soil suction were measured with the solute suction kept approximately constant.

The results in Figs 7(a) and 7(b) clearly show that the Keswick clay (class 1 soil) of the Pleistocene series at depths of 3–4 m exhibits the greatest volume change potential, much more than the surface layer of the black earth profile itself. A value of $I_{p''}$ in excess of 0·10 can be anticipated and when coupled with desiccation by tree roots very large movements can be expected according to equation (2).

The only red-brown earth profile to be tested thus far is the actual profile of the type RB3 quoted by Aitchison *et al.* (1954). This profile, in the grounds of the Waite Research Institute of the

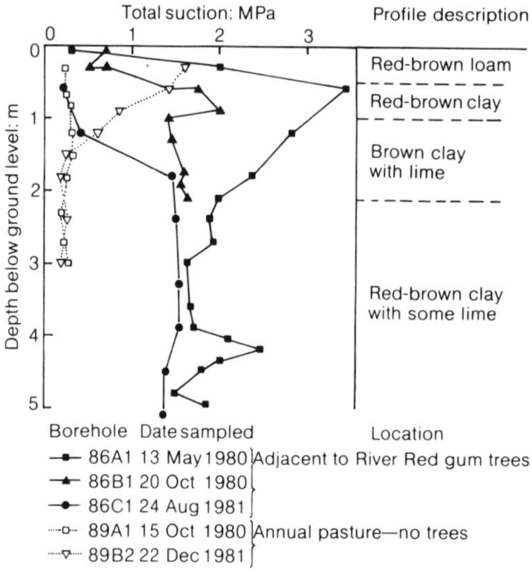

Fig. 9. Measured values of total soil suction in red-brown earth profiles adjacent to River Red gum trees and in an area void of trees

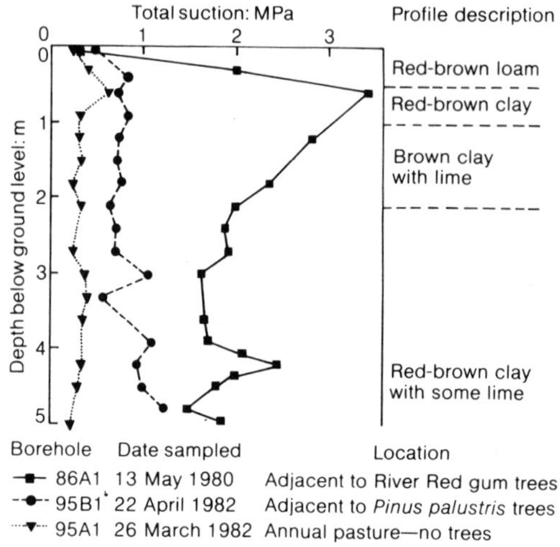

Fig. 10. Measured values of total suction in red-brown earth profiles adjacent to River Red gum trees, next to pine trees and in an area void of trees

University of Adelaide, is the type most commonly found in Adelaide. The soil profile is shown in Fig. 8 together with values of $I_{p''}$ obtained using the first method (see also Emerson & Peter, 1981). Comparison with Figs 7(a) and 7(b) show that the profile is potentially only moderately expansive. The main difference between the two profiles above is whether dispersive clay layers (class 1 soil, e.g. Keswick clay) occur in the top 10 m. In fact, preliminary evidence suggests that the depth to such layers in the profile is the most important soil factor determining the potential for ground movement in the Adelaide area.

Moisture control by vegetation

The first results of the effect of trees were obtained during the sampling programme for the red-brown earth (RB3) profile above. Large River Red gums (*Eucalyptus camaldulensis*) were growing in an open woodland setting permitting soil suction samples to be taken from boreholes, adjacent to a large tree and in an open field nearby. The results, shown in Fig. 9, indicate that the suction varies from about 0·25 to 2·0 MPa. This is equivalent to a change in log suction of 0·9. This difference is assumed to be caused by the trees. Fig. 9 also shows the depth of seasonal variation of moisture to be about 1·4 m under grass and 2·4 m under the gum tree. Under the gum tree, the soil profile remains desiccated all year, except for the wetting of the surface layers during winter rainfall. The limit of penetration of the rainfall was confirmed by measuring the distribution of chloride, introduced by the rain, down the profile. The chloride concentration reached a maximum value of 0·7 meq/100 g soil at 1 m, decreasing rapidly to 0·2 meq/100 g at 3 m. Under the grass site, the chloride values are low with the value at 3 m being 0·07 meq/100 g.

Figure 10 shows three suction profiles measured at approximately the same time of the year, one near the gum tree above, one in the nearby field and the third adjacent to a large pine tree in the same area. These results suggest that the gum tree causes greater desiccation than the similarly sized pine tree at that site.

As a result of this evidence, a larger programme of research was undertaken to study in detail soil moisture conditions adjacent to rows or groups of trees. Some factors being considered are distance from trees, species of trees, soil types and presence or absence of water tables, The monthly rainfall

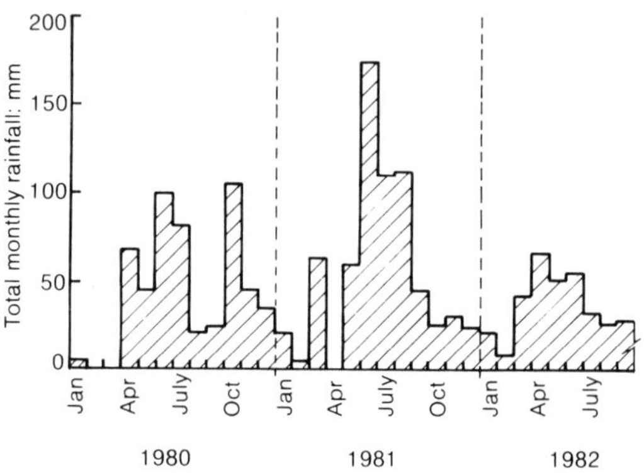

Fig. 11. Monthly rainfall data as recorded in Adelaide from January 1980 to September 1982

data from the beginning of this programme in 1980 to September 1982 are shown in Fig. 11.

One interesting site in the South Parklands provided an opportunity to examine the effect of three groups of different species of large trees. Groups of eucalyptus, casuarina and pine trees occur within a few hundred metres of each other in a natural unirrigated grassed setting on a red-brown earth soil overlying Pleistocene (Keswick) clay. Soil suction measurements show the same trend as above, that large eucalyptus trees have a potential to desiccate the soil profile more than pine trees of similar size. This supports the view that many Australian native species of vegetation are very effective in using soil water, causing rapid desiccation in the dry season. The results from the South Parklands site are shown in Figs 12–14.

A site at Glenside, where a long row of 25 m high pine trees are growing on a red-brown earth, had a water table at a depth of about 6 m. In this case, Keswick clay was not encountered. The results in Fig. 15 indicate that with the ready availability of water there was little desiccation of the profile, except near the surface. This site was later excavated for a large building and the root density was observed in deep trenches. While the root

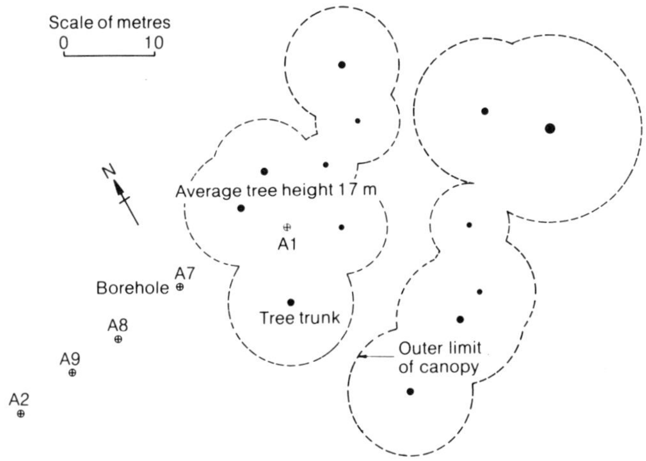

Fig. 12(a). Feature plan of the site surrounding a group of eucalyptus trees in the South Parklands of Adelaide

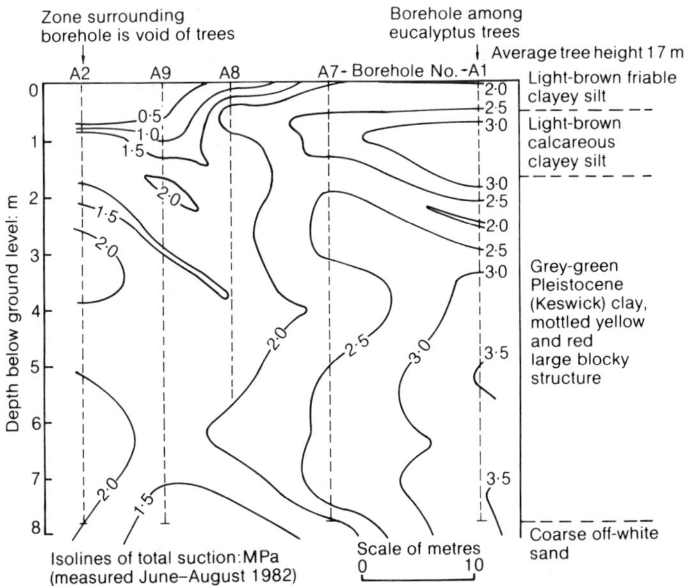

Fig. 12(b). Soil profile and contours of total soil suction near a group of eucalyptus trees in the South Parklands of Adelaide

density was highest near the trees, living roots were detected as far as 47 m from the trees. Roots were also detected at all depths, but a predominance of roots was observed at the junction of the A (loam) and B (heavy clay) horizons where perched water tables exist in the wet season and just above the water table itself.

CONCLUSIONS

Effects of vegetation on ground movements

While much is now being learned about the effects of vegetation on swelling and shrinking soils, not enough knowledge is available to make reliable predictions of ground movements for engineering design purposes. In spite of this several important general conclusions can be proposed at this stage.

First, vegetation can cause significant movements in swelling soils, which can damage engineering structures. The primary effect of vegetation is to cause desiccation of the soil resulting in localized settlement. However, uncontrolled irrigation or watering of gardens may cause the reverse effect resulting in localized heave.

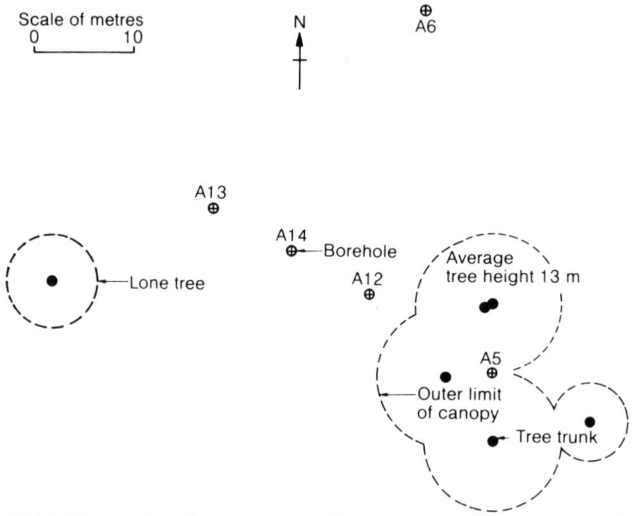

Fig. 13(a). Feature plan of the site surrounding a group of casuarina trees in the South Parklands of Adelaide

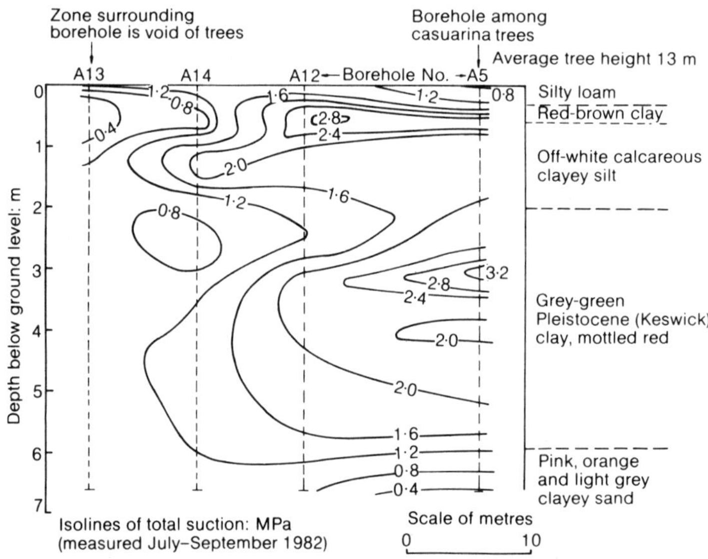

Fig. 13(b). Soil profile and contours of total soil suction near a group of casuarina trees in the South Parklands of Adelaide

Studies of soil moisture conditions, ground movements and physical evidence of roots have shown that vegetation can influence the soil under covered areas to a much greater extent than variation in climate alone. In Adelaide, these effects have been measured to depths in the expansive clays, in excess of 10 m, and distances of 47 m from large trees, greater than 1·5 times their height.

However, the magnitude of the effects of vegetation is not simple and is dependent of many factors, e.g. size and species of tree, soil type, hydrology. The studies undertaken in Adelaide, while not yet conclusive, can provide some guidelines. Native species of trees cause greater desiccation than the exotic trees examined, with eucalyptus species having a greater effect than the casuarina species. The upper limit of total suction due to desiccation for eucalyptus species in Adelaide is in excess of 4 MPa. The lower limit of total suction due to excessive irrigation is a function of soil salinity, but is rarely less than 0·3 MPa in Adelaide.

The effect of vegetation on total soil suction is

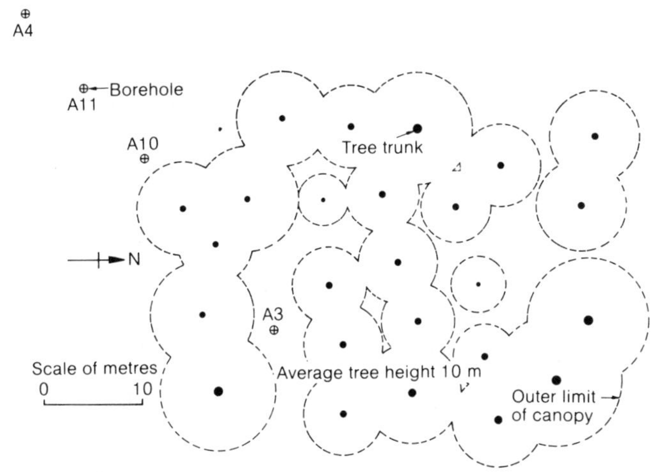

Fig. 14(a). Feature plan of the site surrounding a group of pine trees in the South Parklands of Adelaide

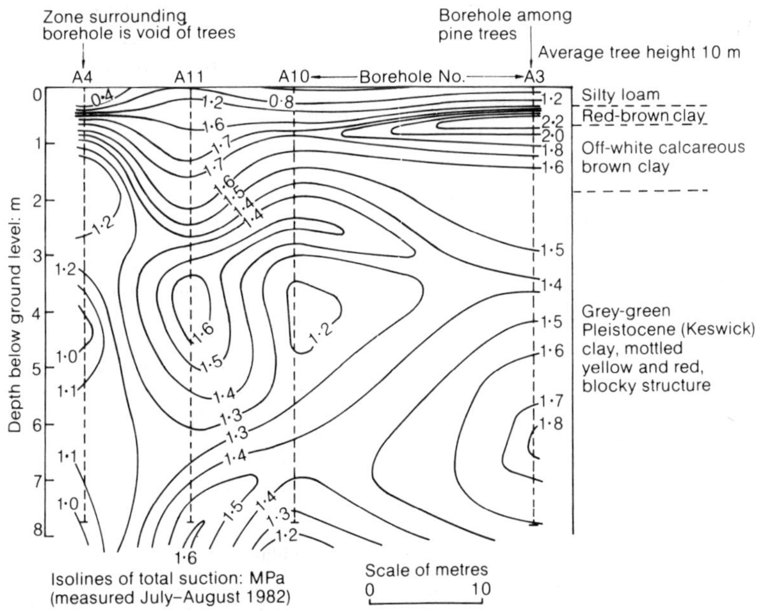

Fig. 14(b). Soil profile and contours of total soil suction near a group of pine trees in the South Parklands of Adelaide

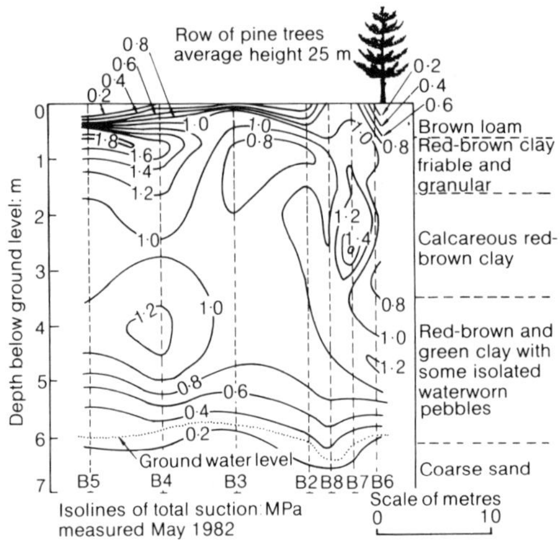

Fig. 15. Soil profile and contours of total soil suction near a row of tall pine trees on a red-brown earth at Glenside, South Australia

much greater in the less expansive clays, e.g. red-brown earths than the highly expansive clays, e.g. black earths and the underlying Pleistocene clays. This is due to several reasons:

(*a*) the lower degree of saturation, i.e. better aeration, higher permeability and availability of water below wilting point and lower salinity of the less expansive clays, assists root development

(*b*) the lower salinity found in less expansive clays results in lower total suctions in situ permitting larger changes in suction to occur before wilting point is reached

(*c*) the higher permeability of the less expansive clays permits changes in soil suction to occur more rapidly.

While the total suction changes in the highly expansive clays may be less, this does not mean that the ground movements will be less, as values of $I_{p''}$ for such soils are much higher. In fact, ground movements well in excess of 150 mm due to trees have been recorded in these soils in Adelaide.

Finally, the presence of a permanent shallow water table of low salinity will reduce soil movement by deep rooted vegetation. However, changing the availability of water to vegetation, for example by altering the site watering pattern or drainage, may lead to excessive soil movement.

Practical design methods

Various analytical methods have been proposed for the design of light structures on expansive clays, e.g. Lytton (1970), Walsh (1978) and Mitchell (1979). The most significant factors in the application of these methods is the estimation of the unloaded surface movement, defined as y_m, and the shape of the ground surface under the structure due to this movement and defined in various ways. The use of an equation of the form of equation (2) has been used in most cases. However, the Authors' experience and that of Holland & Cameron (1981) suggests that the practical use of equation (2) is unreliable at the present time and in fact has not always been able to predict even the direction of the movement. This is particularly true where vegetation and its management such as watering is concerned. The main problems with the use of equation (2) are that first $I_{p''}$, expressed as a simple constant, cannot truly represent a material as complex as expansive clay and, secondly, the suction change cannot be estimated accurately with our current knowledge. It is mainly for these reasons that these analytical methods of design have not proved successful in Australia.

Perhaps the most successful design procedure used in Australia is that used in the Melbourne area (Walsh, Holland & Kouzmin, 1977). This does not use estimates of ground movements nor analytical methods directly, but correlations with classified soil conditions based on wide experience and knowledge gained through use of the analytical methods. This may suggest that ultimately probabilistic design methods based on correlations with geomorphological classifications may be the best solution. Analytical procedures would be a necessary part of these methods, particularly for comparing alternate designs. However, one of the important problems still to be resolved is whether one should design for the most severe case experienced in Australia, that is a large eucalyptus tree adjacent to a house on deep expansive clay. The final decision on this problem will depend on socio-economic and legal considerations.

REFERENCES

Aitchison, G. D. (1965). Editor: *Moisture equilibria and moisture changes in soil beneath covered areas*, pp. 1–278. Sydney: Butterworths.

Aitchison, G. D. (1973). The quantitative description of the stress–deformation behaviour of expansive soils. *Proc. 3rd Int. Conf. Expansive Soils, Haifa* **2**, 79–82.

Aitchison, G. D. & Holmes, J. W. (1953a). Seasonal changes of soil moisture in a red-brown earth and a black earth in Southern Australia. *Aust. J. Appl. Sci.* **4**, No. 2, 260–273.

Aitchison, G. D. & Holmes, J. W. (1953b). Aspects of swelling in the soil profile. *Aust. J. Appl. Sci.* **4**, No. 2, 244–259.

Aitchison, G. D., Peter, P. & Martin, R. (1973). The instability indices I_{p_m} and I_{p_s} in expansive soils. *Proc. 3rd Int. Conf. Expansive Soils, Haifa* **2**, 101–104.

Aitchison, G. D., Peter, P. & Power, R. F. (1977). Stress–deformation mechanisms in clays as determinants of design requirements for small and large structures—

an evaluation through case histories in Adelaide of relative significance of several mechanisms. In Engineering conference, Mini-symposium *How to live with geotechnical risks, Cooma, New South Wales*, 26–33.

Aitchison, G. D. & Richards, B. G. (1965). A broad scale study of moisture conditions in pavement subgrades throughout Australia. In *Moisture equilibria and moisture changes in soil beneath covered areas*, pp. 184–232. Sydney: Butterworths.

Aitchison, G. D., Sprigg, R. C. & Cochrane, G. W. (1954). *The soils and geology of Adelaide and suburbs*. Bulletin No. 32. Adelaide: Geological survey of South Australia.

Emerson, W. W. & Peter, P. (1981). The engineering properties of red-brown earths. Prescott Symposium on *Red-brown earths of Australia*, pp. 149–163. Adelaide: Commonwealth Scientific Industrial Research Organization.

Holland, J. E. & Cameron, D. A. (1981). Seasonal heave of clay soils. *Civ. Engng Trans. Instn Engrs Aust.* CE23, No. 1, 55–67.

Lytton, R. L. (1970). *Design criteria for residential slabs and grillage rafts on reactive clay*. Technical Report. Melbourne: Commonwealth Scientific Industrial Research Organization.

Mitchell (1979). *The structural analysis of footings on expansive soil*. Research Report No. 1. Australia: K. W. Smith & Associates.

Norrish, K. & Pickering, J. (1977). Clay mineralogic properties. In *Soil fabric in crop production in a semi-arid environment* (ed. J. S. Russell & E. L. Graecen), chap. 3. Brisbane: University of Queensland Press.

Northcote, K. (1974). *A factual key for the recognition of Australian Soils*, p. 3. Adelaide: Rellim Technical Publications.

Northcote, K., Hubble, G. D., Isbell, R. F., Thompson, C. H. & Bettenay, E. (1975). *A description of Australian soils*. Melbourne: Commonwealth Scientific Industrial Research Organization.

Peter, P. & Martin, R. (1973). A simple psychrometer for routine determinations of total suction in expansive soils. *Proc. 3rd Int. Conf. Expansive Soils, Haifa* **2**, 89–76.

Richards, B. G. (1965). Measurement of free energy of soil moisture by the psychrometric technique using thermistors. In *Moisture equilibria and moisture changes in soil beneath covered areas* (ed. G. D. Aitchison), pp. 39–46. Sydney: Butterworths.

Richards, B. G. (1967). Moisture flow and equilibria in unsaturated soils for shallow foundations. *Permeability and Capillarity*, ASTM Spec. Techn. Publ. 417, pp. 4–34. Philadelphia: American Society for Testing and Materials.

Richards, L. A. & Weaver, L. R. (1943). 15 atmosphere percentage as related to the permanent wilting percentage. *Soil Sci.* **56**, No. 5, 331–339.

Russam, K. & Coleman, J. D. (1961). The effect of climatic factors on subgrade moisture conditions. *Géotechnique* **11**, No. 1, 22–28.

Schumann, E. E. (1967). The effect of environment on soil moisture suction. *Proc. 5th Aust.-N.Z. Conf. Soil Mech., Auckland* **1**, 101–106.

Selby, J. (1979). Soil classification and mapping. In *Footings and foundations for small buildings in arid climates (with special reference to South Australia)*, pp. 20–31. Adelaide: I. E. Australia.

Stace, H. C. T., Hubble, G. D., Brewer, R., Northcote, K. H., Sleeman, J. R., Mulcahy, M. J. & Hallsworth, E. G. (1968). *A handbook of Australian soils*. Glenside, South Australia: Rellim Technical Publications.

Standards Association of Australia (1980). Determination of Emerson class number of a soil. In *Methods of testing soils for engineering purposes*. AS 1289.C8.1. Australia: Standards Association of Australia.

Stapleton, D. H. (1970). Changes and some structural defects developed in some South Australian clays and their engineering significance. *Proc. Conf. Soils Earth Structures in Arid Climates, Adelaide* **1**, 62–71.

Taylor, J. K., Thompson, B. P. & Shepherd, R. G. (1974). *The soils and geology of the Adelaide area*. Department of Mines, Bulletin No. 46. Adelaide: Geological Survey of South Australia.

Walsh, P. F. (1978). *The design of residential slabs-on-ground* (Revised method). Technical Paper No. 25. Melbourne: Commonwealth Scientific Industrial Research Organization, Division of Building Research.

Walsh, P. F., Holland, D. E. & Kouzmin, T. (1977). *The classification of the expansive behaviour of Melbourne soils for domestic construction*, pp. 1–11. CSIRO and AEBIRA Report. Melbourne: Cement and Concrete Association of Australia.

Ward, W. H. (1953). Soil movement and weather. *Proc. 3rd Int. Conf. Soil Mech., Zurich* **2**, 477–481.

Willoughby, D. R. (1965). Groundwater observations during the initial period of urban development at Elizabeth, South Australia. *Proc. 1st Int. Conf. Expansive Clay Soils, Texas*, 31–44.

Evapo-transpiration and heaving clays in South Africa

A. A. B. WILLIAMS* and J. T. PIDGEON*

Examples are given of the depth of root systems in the highveld. The heaving clays found in this region can cause extensive damage to buildings. The ranges of soil suction are often in excess of pF 4·2; and in situ methods of measuring it are discussed. The removal of vegetation causes the most significant change in the stress state of the soil, which exceeds any changes which may be caused by building load or excavation for small structures.

Des exemples sont donnés de la profondeur des systèmes de racines dans le highveld. Les argiles gonflantes qui se trouvent dans cette région peuvent produire des dégâts considérables aux bâtiments. Les chiffres pour la succion caractéristique des sols ici dépassent souvent pF 4·2. Des méthodes de mesurer celle-ci in situ sont discutées. C'est l'enlèvement de la végétation qui produit le changement le plus important dans l'état de contrainte du sol. Ce changement qui en lui-même est plus important que n'importe quel changement provoqué soit par une excavation soit par la charge induite par une petite structure.

INTRODUCTION

The occurrence of heaving clays is widespread throughout South Africa due to a unique combination of features which also affect their origin and behaviour

(*a*) the distribution of source rocks, of igneous or sedimentary type, whose composition results largely in the formation of montmorillonitic materials on weathering: thus active residual soils are formed in situ or active transported soils occur in the form of clays in fine colluvium, gulley-wash, alluvium or lacustrine deposits; glacial clays do not exist as is so often found in the northern hemisphere and usually affected by solifluction

(*b*) the geological age of the land surfaces: in some areas relicts of the African geomorphic cycle, beginning in mid-Cretaceous times and lasting more than 30 million years, are found (Brink, 1979)

(*c*) the type of climate and its duration in much the present form for about one million years: most of the interior, or highveld, has no surplus moisture in any season so that the soils are in a desiccated state for most of the year (Fig. 1). Further, a particular type of vegetation has developed and endured for a long period.

Without human interference these elements would be in a state of dynamic equilibrium with time cycles of between one and several hundred years. The combination of these factors results in a specific type of vegetation which can indicate the degree of similarity of the environmental factors to those in other countries.

PRACTICAL PROBLEMS AND OCCURRENCE OF ROOTS

The desiccating effect of trees is known to cause active clays to shrink and this can cause damage to buildings. However, few detailed observations of the magnitude of the movements involved have been made. Donaldson & Kleynhans (1963) monitored the movements of a house in Vereeniging for 10 years. The area is notorious for heaving clay and one point on the western wall reached a heave of 80 mm over a five year period, after which only a very slight seasonal variation of a millimetre or two occurred. On the south side of the house peach and apricot trees had been planted at the beginning of the monitoring period, about 5 m away from the wall. After an initial heave of about 72 mm, the displacement decreased to 22 mm. The active lacustrine clay extends 1·5–8 m below the surface and the trees clearly caused a shrinkage settlement of about 58 mm.

In another instance, a relative movement of 200 mm was recorded on a house at Allanridge, in the Orange Free State (Fig. 2). The movement was attributed to the desiccating effect of a row of blue gum trees adjacent to the building, on an expansive residual soil layer located 5 m below the surface and exceeding 8 m in thickness (Pidgeon & Tromp, 1983).

These are typical examples for the region, and the behaviour of shrinkable clays in the UK, Canada and South Australia is similar. However, it is the swelling of heaving clay which causes most damage in this country and in areas of the southern USA, Israel, northern Australia and India. This study indicates that it is the initial desiccating effect of natural vegetation (by transpiration) rather than any later evapo-transpiration from the soil which causes most of the trouble on the South African highveld where active layers exist at some depth. However, there are cases where problems due to shrinkage arise under certain combinations of soil profile type and conditions existing at the time of construction. The

* National Building Research Institute, South Africa.

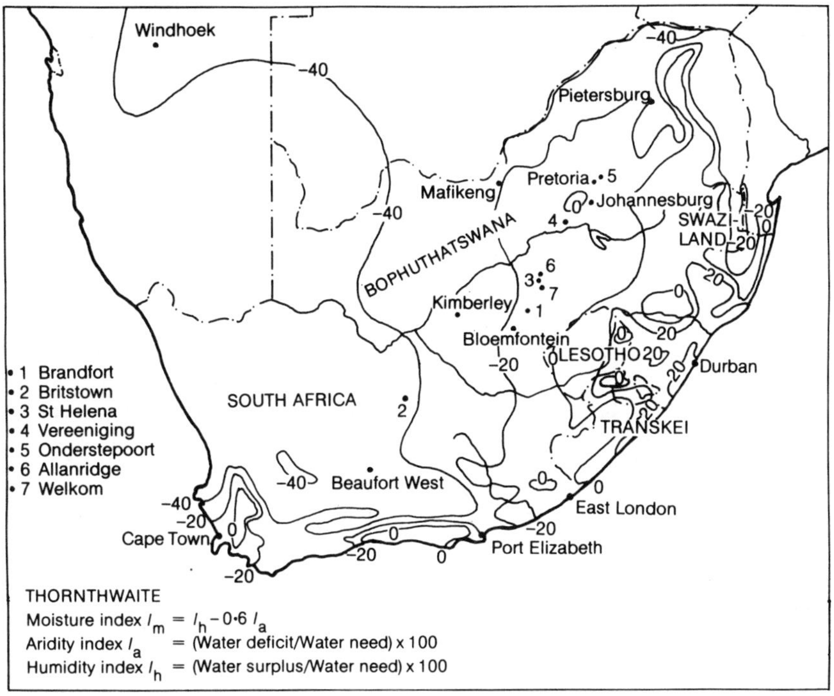

Fig. 1. Contours of Thornthwaite moisture index in South Africa

moisture regime in an unsaturated soil is best considered in terms of the degree of soil suction over the whole soil profile, using the pF* scale.

Early work by Wolpert (1952) on a very deep active residual clay profile at the Council for Scientific and Industrial Research (CSIR) site in Pretoria revealed that soil suction values were very much higher than expected from a linear extrapolation of the hydrostatic condition below the water table, which was at a depth of about 30·5 m. At a depth of 6 m, the pF was 4·2 and at a depth of 15 m the pF was 3·8; under hydrostatic conditions these readings would have been 3·4 and 3·2 respectively. A similar case of a very deep desiccated profile was reported by Williams (1980) in Kimberley; the very stiff laminated residual clay exhibited total suction values of pF 4·6 and were generally above 4·2 (commonly regarded as the wilting point of vegetation) to a depth of 13 m. Fig. 3 shows the measured values compared with the extrapolated hydrostatic condition. The presence of gypsum in the profile may have introduced a solute suction component.

Little is known about the actual distribution of roots below any vegetative cover but, in the case of trees, an approximation of their extent is given by the network of branches and twigs above the ground. This varies with the type of tree, as mentioned in a survey of damage caused by trees and shrubs in the UK by Cutler & Richardson (1981). They found that damage to buildings could occur up to 40 m from a tree; which gives an indication of the lateral extent of the root system. Regarding the depth of occurrence, roots beneath a gum tree (eucalyptus) plantation on the CSIR site were found at 27·5 m (Wolpert, 1952) which compares with a depth of 30 m for eucalyptus in Australia mentioned by Jacobs (1955). These are isolated records, but a review of common profiles in South Africa for which detailed descriptions do exist indicates that easily visible roots seldom extend below half the depth to the water table (WT). Roots at Kimberley extended to 5 m (WT below 30 m); at Onderstepoort to 2·5 m (WT below rock which is at 4 m), and at Vereeniging to 3·5 m (WT at 8 m). It is assumed that the roots of vegetation in the highveld region would not be found below the water table, as aeration is required for root growth. This feature is used as an indicator in the engineering description of soil profiles (Jennings, Brink & Williams, 1973) when the water table conditions are obscure.

THE DESICCATING EFFECT OF ROOTS

Transpiration from vegetative cover is known to be the major cause of loss of water from soils in semi-arid regions; reviews of some of the implica-

* pF = $\log_{10} h$ (h is the height of water in cm).

tions of this for buildings and roads in South Africa have been given by Williams & Simons (1963) and de Bruijn (1973a). The most significant change in loading made to a building site, i.e. to the stress state in the soil, is considered to be the removal of vegetation—all other effects, such as imposed foundation loads, drainage or lowering of the ground water table, or relief of stresses on excavation are of secondary importance. The distribution of soil suction in a profile is controlled primarily by evapo-transpiration from the vegetation and, when this is removed, because the influence of evaporation is minimal and confined to a shallow layer near the surface, soil moisture will accumulate after rain (de Bruijn, 1973a). The maximum suction below this layer is controlled by the ability of the plants to extract moisture from the soil, and the level at which moisture ceases to be available to the plant is called the permanent wilting point. This value varies for different soils but generally corresponds to a negative matric potential of pF 4·2 in the case of crops. In climates with distinct wet and dry seasons, the indigenous vegetation has evolved a drought tolerance in order to survive, which enables it to use soil water at a matric potential of pF 4·2 or more over long periods. Where the effects of evaporation are felt deeply into the main root zone, the plant must be able to prevent moisture being withdrawn through the roots. This situation is only likely to occur with shallow rooted trees and shrubs, and with grasses which have been burned in the drier months (to encourage early grazing with the first rains). Any tendency for the ground surface to be shaded either by the plant itself or by larger trees will reduce evaporation while the cracking of the soil mass will provide a larger surface area for evaporation. For example, if the soil is covered with live vegetation the suction will be approaching the wilting point of pF 4·2 even during the highveld summer rainfall season. This is equivalent to a negative pressure of 1555 kPa and could be considered equivalent to the maximum swelling pressure which a clay could exert. If the pF is 4·5, the stress state in the soil would be equivalent to an over-consolidation pressure of about 3100 kPa.

The primary significance of vegetation was indicated by de Bruijn (1965) when he reported on a site in the vicinity of Odendaalsrus (Freddies South gold mine), noted for cracked buildings caused by heaving clay. Under a surface cover of trees or natural grass the moisture content of the stiff, grey mottled orange, sandy clay was only 13% (corresponding to an in situ pF of 4·0) throughout a layer 2–5·5 m deep. Under a bare gravel surface, which had been used as a parking ground for many years, the moisture content in this layer was 17% (corresponding to a pF of 3·4).

De Bruijn (1973a) also reported on an experimental site on black expansive clay at Onderstepoort. The difference in swelling and shrinking of a naturally vegetated area subject to transpiration loss was compared to a similar area of bare ground where the surface had been treated with defoliant. The records of seven years of precise

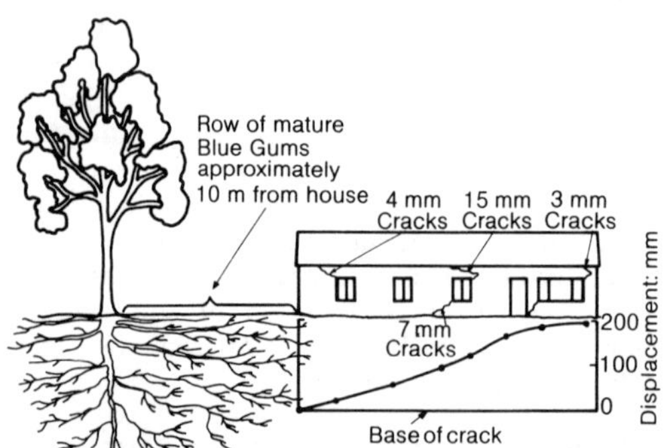

Fig. 2. Level survey of plinth of house in Allanridge

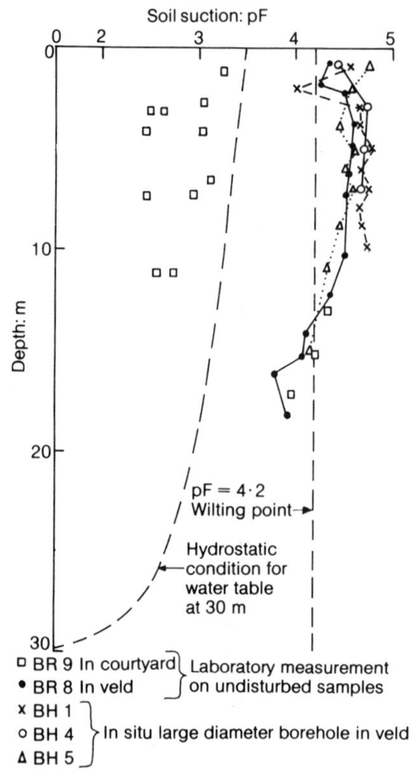

Fig. 3. Psychrometer readings of soil suction near Kimberley

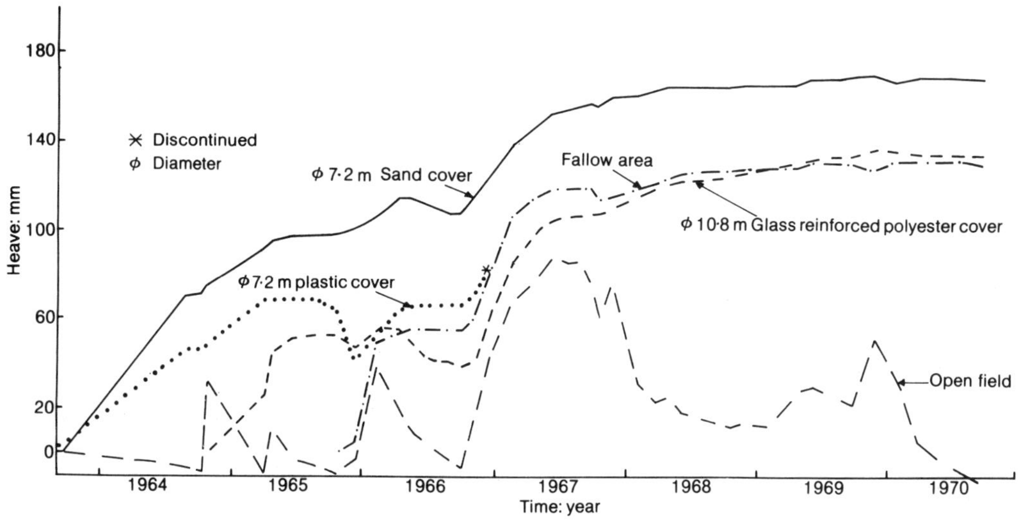

Fig. 4. Heave against time curves for Onderstepoort test sites

level observation are shown in Fig. 4 and the conclusions are

(*a*) The level of the fallow area, or bare ground with clay up to the surface, heaved to about 130 mm and reached a fairly stable equilibrium. (The maximum heave recorded on this site was about 170 mm. This occurred under a blanket of clean sand and stabilized to within a few mm.)

(*b*) The level of the natural grass area showed an irregular but cyclical seasonal movement with an amplitude of 50–60 mm (the 100 mm amplitude in 1967 was due to an unusually humid season).

It has been found in more recent investigations at Onderstepoort that at depths below 1 m, where direct evaporation is considered to have little effect, suctions of up to pF 4·4 exist due to transpiration effects (see Fig. 5). Wide vertical cracks exist to depths of 600 mm in this profile, and the suction associated with the onset of cracking is approximately pF 4·6. During the wet season the suctions may drop to

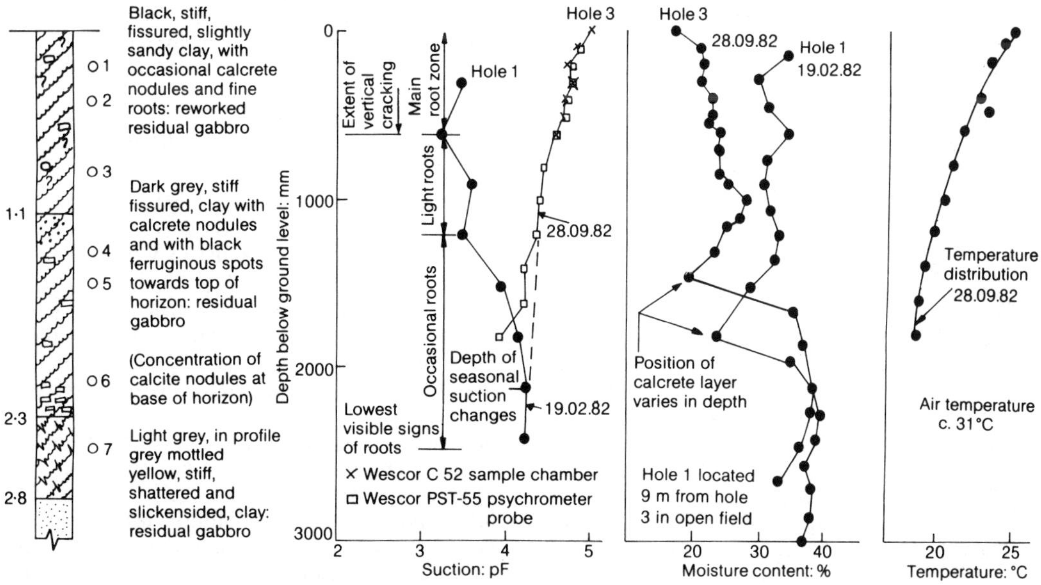

Fig. 5. Seasonal variation of the suction and moisture content profiles at Onderstepoort

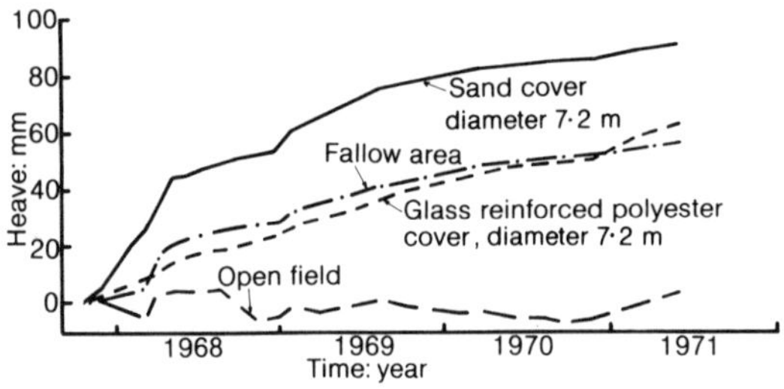

Fig. 6. Heave against time curves for Vereeniging test sites

Table 1. Effect of plant type on transpiration

Type of plant	Transpiration of moisture on a sunny day
Gum tree (*Eucalyptus macarthuri*)	500 l/day
Black wattle (*Acacia mollissima*)	250 l/day
Highveld grass (*Themeda*)	1 l/m^2/day

approximately pF 3·2 and suction changes can be expected over a depth of 2·1 m between wet and dry seasons. The very high suctions attributed to the grasses at the Onderstepoort test site are not unique for plants as a whole. Holland (1982) states that values of up to pF 4·4 are not uncommon for certain types of trees, Pile (1981) quotes a figure of pF 4·49 for Australian eucalyptus and Mitchell (1982) mentions typical values of pF 4·5 for trees in Australia.

De Bruijn (1973b) conducted another experiment, similar to that at Onderstepoort, on the swelling and shrinking of Vereeniging clay. The results are reproduced in Fig. 6 which covers a 4 year observation period.

(*a*) The ground level of the area which had been laid bare heaved to a fairly constant 60 mm (in comparison with the maximum of 90 mm under an artificial sand blanket).

(*b*) The ground level of the natural grass area showed a seasonal movement with an amplitude of 12 mm. (The smaller seasonal variation, in comparison with Onderstepoort, is because of the type of soil and the composition of the soil profile.)

Further, observations on seasonal movements of natural ground were made by Williams & Simons (1963) at three sites: Brandfort, Britstown and St Helena. Seasonal movements in these areas were between +10 mm and −10 mm; and the possible heave of covered surfaces was known to be about 55 mm. Thus two further important conclusions can be drawn

(*a*) The full depth of active clay in the profiles is unlikely to become fully wet up during the wet season (or a much greater amplitude of movement would be observed). The soil suction in the profile is apparently maintained by the transpiration of vegetation, even during the summer growth period.

(*b*) No relationship appears to exist between the seasonal movement occurring under natural conditions and the total swell which results when the surface is covered by a structure. The total depth of active material is unaffected by seasonal and shallow-seated moisture changes, and is to be expected. In this region of semi-arid conditions with deep water tables (10–30 m or more) the concept of the depth of seasonal movement cannot be applied to foundations for buildings or large covered areas. Such a concept may be applicable where the natural regime is not disturbed much, as could be the case for pylon foundations or poles for overhead services.

The swelling movements on heaving clays in South Africa associated with the removal of vegetation and the imposition of any subsequent structure or surface treatment, amount to about 150 mm in many areas, and the most extensive heave recorded to date (at the site in Kimberley) is 374 mm. In a typical highveld regime the mean annual precipitation is about 700 mm (occurring during the summer months between October and April), and the potential evapo-transpiration is about 800 mm. The deficiency is therefore in the region of 100 mm; Thornthwaite's aridity index (*Ia*) is 15, and the moisture index (*Im*) −9 (Fig. 1). The transpiration from vegetation depends on many factors, but an example of the extent is given by Henrici (1946, 1947)—see Table 1.

DESIGNING FOR THE EFFECTS OF VEGETATION

Knowledge of a particular plant's ability to survive at any site, its suction characteristics and depth of root zone provides useful design guidance on the highest levels of suction likely to occur on the site. This information is important because the site investigation for a project may take place several months or years prior to construction, and suction measurements taken during the investigation may bear no resemblance to those existing at the time of construction. Heave prediction should be based on an assessment of the suction profile which would exist at the time of construction, and the extreme dry profile condition would provide the most conservative estimate. The design must allow for the development and management of the vegetation to be established after construction, either the re-establishment of indigenous vegetation or the introduction of exotic plants, because of varying water demand and/or suction characteristics.

These effects will influence the structure in different ways depending on the type of soil profile and the length and intensity of any unusual climatic features. Because many developed areas of South Africa are approaching the limit of their water resources, severe restrictions on the use of water for gardening should be borne in mind.

If an attempt is to be made to quantify the influence of pre-construction or post-construction vegetation on the volume change behaviour of an expansive soil profile, consideration must be given to both the water demand and permanent wilting point characteristics of the vegetation, and the moisture characteristic (moisture content plotted against suction relationship) of the soil. The latter should be determined preferably from in situ measurements, although in the upper self mulching portion of the Onderstepoort profile, good correspondence has been obtained between in situ measurements using Wescor thermocouple psychrometers and results obtained by Brackley (1975) on compacted remoulded soil (see Fig. 7). The moisture characteristic of the soil yields valuable information regarding the moisture contents corresponding to the field capacity (defined in terms of suction, pF 2·0) and the permanent wilting point (pF 4·2–4·5), and the rate at which the suction changes with variation of moisture content. This enables an assessment to be made of the range of suctions and moisture contents likely to occur in the zone of seasonal variation. The values below this zone may be determined by direct measurement in situ. The extent to which the vegetation is capable of

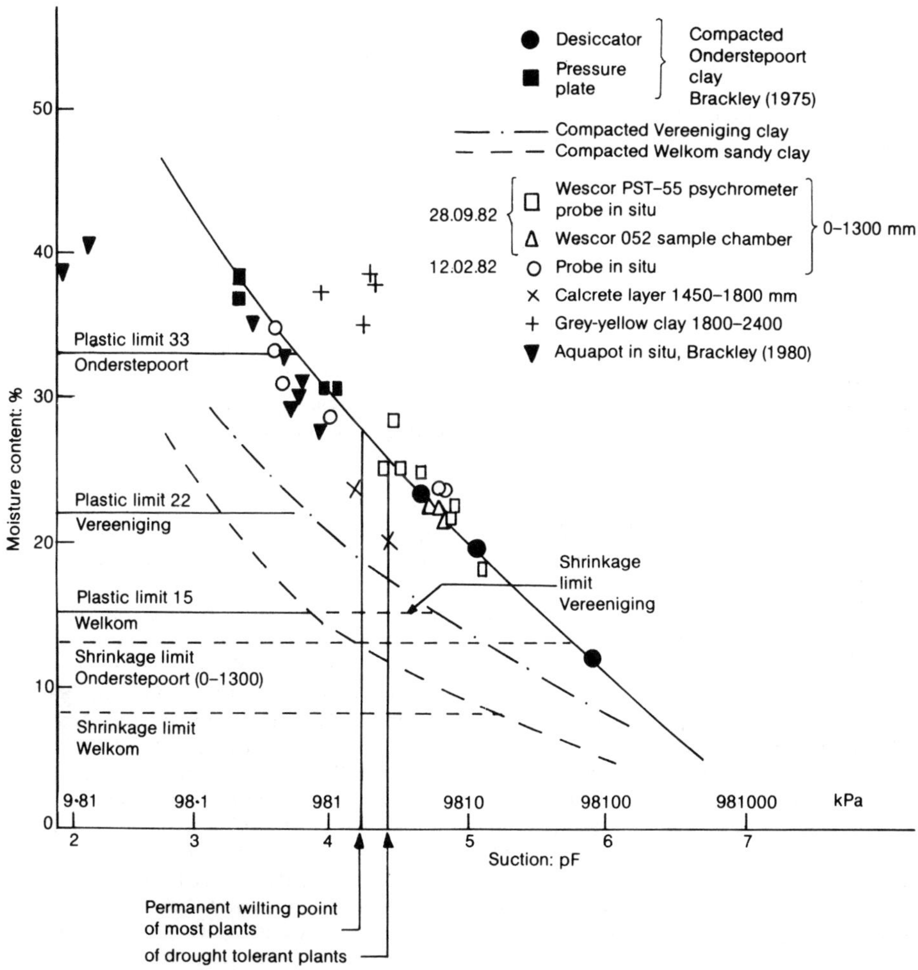

Fig. 7. Moisture content suction relationships: Onderstepoort, Vereeniging and Welkom clays

increasing the suction to the level associated with the shrinkage limit of the soil in which it is growing is indicated in Fig. 7. Archer (1975) noted that the moisture content at the wilting point (corresponding to pF 4·2) was greater than the shrinkage limit in soils of high clay content and less in those of low clay content. Fig. 7 supports this observation by indicating that suction at the wilting point is much closer to that at the shrinkage limit for soils of lower clay content. This could explain the observation that soil settlement resulting from the dehydration effects of trees is more pronounced in low to moderately expansive clay areas than in highly expansive areas (Holland, 1982).

The volume change behaviour of the soil can be predicted either from the in situ dry density (or specific volume) or moisture content relationship which also provides useful information on the volume of moisture stored in the soil mass at any particular moisture content; or by adopting a stress state variable approach over the anticipated range of suction changes. The latter approach was followed by Brackley (1980) who held that the maximum heave under a building can be predicted by the equation

$$\text{Swell } (\%) = \frac{(\text{PI} - 10)}{10} \log_{10} \frac{S}{p} \qquad (1)$$

where PI is the plasticity index of the whole sample, S is the soil suction at the time of construction in each layer in kPa and p is the overburden plus foundation pressure acting on each layer in kPa.

This is an empirical relationship based on the measurement of soil suction in situ and in the laboratory, and observations of corresponding heave in the field. If suction changes at various positions under a building can be predicted reasonably accurately from the time the foundations are laid, total and differential heave can be determined using relationships such as given in equation (1).

Volume change due to shrinkage of the upper layers of the profile caused by evaporation is equally difficult to predict accurately. Fortunately most soil profiles on the Transvaal highveld are covered by an inactive surface layer, and others have substantial grass cover which, if not destroyed by customary burning, would shade the soil surface; the effects of evaporation are thus in many instances insignificant. They are in any event likely to be small at the founding level of most structures. If volume change due to evaporation is considered to be important this can be conservatively predicted by assuming the lower limit of the soil moisture content to be the shrinkage limit, as desiccation below this value would not cause further volume change.

METHODS OF MEASURING DESICCATION

It has been established that one of the crucial parameters in gauging the influence of vegetation is the soil suction. There are many methods which have been used to determine soil suction in the laboratory; however, only those methods used in field experiments are described.

Simple tensiometers, filled with air-free water, were used at the Onderstepoort test site with little success. For a few months a perched water table may have existed, but thereafter the measuring range (up to pF 2·9 or −8 kPa) was exceeded. A fair degree of success was achieved using gypsum block field tensiometers (de Bruijn, 1973a). A number of commercially available cylindrical gypsum blocks, 50 mm in diameter, were assembled in stacks, together with temperature sensors and spacers. They were inserted into boreholes and the annular space was filled with kaolinite to act as an interface. Careful calibration was required and although many of the devices gave readings of resistivity for years, their reliability was never proved.

They were general indicators of change and of the relative performance between adjacent installations, rather than absolute gauges. An advantage was the wide range of potential operation, covering suction values between pF 2·8 and pF 4·5. A type of osmotic tensiometer, called an Aquapot, was successfully used at Onderstepoort and Vosloo Park by Brackley (1975) and some results are included in Fig. 7. However, variable zero drift occurred over a period of months, which made the Aquapot unsuitable for long-term monitoring.

During the past few years attempts have been made to develop a field technique for the installation of thermocouple psychrometers, following rapid advances in electronic design and commercial availability. Initial field trials involving psychrometers with ceramic shields were only partially successful, because of the sporadic inundation of the device or corrosion of the exposed thermocouple wires. Severe problems due to this were reported by Daniel, Hamilton & Olson (1979). A new type of psychrometer tip in stainless steel mesh has recently been developed and with the embodiment of a few simple protective measures a more long-lasting in situ suction measuring device seems to have been found. The operation of in situ psychrometers in arid environments for a period of nearly 4 years has been reported in Utah by Wiebe, Campbell, Gardner, Rawlins, Cary & Brown (1971). Details of the National Building Research Institute's present technique, which is modelled on that used in the state of Utah, are given in Appendix 1. (Results using these multiple tip probes have already been given.)

An alternative method for rapid in situ measurement of soil suctions during the site investigation phase of a number of projects involves psychrometer probes placed at intervals down the side of a 750 mm dia. borehole. The use of these large diameter boreholes for site investigation and soil profiling is common in South Africa and it is a relatively simple matter to seal a small probe in a hole in the sidewall. The small holes typically extend 50 mm from the wall of the borehole and have a diameter equal to that of the body of the probe (approx. 8 mm). They are formed using a hand drill with a twist bit and any suitable mouldable plastic material may be used for sealing the hole behind the probe. In the fairly high pF ranges encountered, two hours is usually adequate for equilibrium to be established. While the measurement is not entirely under field stress conditions it is considered to provide a more realistic method of in situ sensing than a laboratory method which is subject to some sample disturbance as well as full relief of stress, and problems associated with maintaining a stable temperature regime are virtually eliminated.

CONCLUSION

The influence of vegetation, and its removal, on potentially expansive or shrinkable soils depends on the climate, the type of vegetation, the types of soil and their position in the soil profile and the location of either the water table (if one exists) or an impermeable lower boundary. In the natural state vegetation can survive only if its environment is conducive to growth. It will interact with the environment to create a condition of dynamic equilibrium when viewed on an extended time scale. Seasonal and yearly fluctuations in this condition may cause buildings constructed at different times to perform quite differently. Suction measurements taken in situ at various locations in the semi-arid and sub-humid areas of South Africa, where water tables do exist, have revealed that soil suctions considerably higher than those corresponding to simple hydrostatic conditions exist right down to the water table. The effect that this has on buildings will depend among other things on their size in plan relative to the depth to water table.

The measurement of suction in situ using commercially available thermocouple psychrometers has been found to be feasible both for rapid results during the site investigation phase and for long-term monitoring. More information is required on the seasonal fluctuation of soil suctions in the natural state, the lateral variation under buildings and the limiting values of soil suction (corresponding to the permanent wilting point) of various types of vegetation. The latter, together with the water demand, when compared with the suction corresponding to the shrinkage limit of the soil and the amount of water held in the soil mass at a particular moisture content, indicates the potential of the vegetation to create volume changes in the soil under extreme climatic conditions. The Thornthwaite moisture index is not an adequate indicator of climate for this purpose and the shrinkage limit should preferably be obtained on undisturbed samples. The measurement of seasonal heave under natural conditions may give very little indication of the potential for total or differential heave under a building.

REFERENCES

Archer, J. R. (1975). Soil consistency. In *Soil physical conditions and crop production*. Technical Bulletin 29, pp. 289–297. London: Ministry of Agriculture, Fisheries and Food.

Brackley, I. J. A. (1975). *The interrelationship of the factors affecting heave of an expansive, unsaturated, clay soil*, pp. 159–166. PhD thesis, University of Natal.

Brackley, I. J. A. (1980). Prediction of soil heave from suction measurements. *Proc. 7th Reg. Afr. Conf. Soil Mech., Accra* **1**, 159–166.

Brink, A. B. A. (1979). *Engineering geology of Southern Africa*, Vol. 1. Pretoria: Building Publications.

Cutler, D. F. & Richardson, I. B. K. (1981). *Tree roots and buildings*. London: Construction Press.

Daniel, D. E., Hamilton, J. M. & Olson, R. E. (1979). Suitability of thermocouple psychrometers for studying moisture movement in unsaturated soils. *Proc. Symp. A.S.T.M., Pennsylvania.*

de Bruijn, C. M. A. (1965). Some observations on soil moisture conditions beneath and adjacent to tarred roads and other surface treatments in South Africa. NBRI R/Bou 155. pp 135–142. Pretoria: Council for Scientific and Industrial Research.

de Bruijn, C. M. A. (1973a). Moisture redistribution in Southern African soils. *Proc. 8th Int. Conf. Soil Mech., Moscow* **2**, 37–44. (NBRI R/Bou 422).

de Bruijn, C. M. A. (1973b). Moisture redistribution and soil movements at Vereeniging. *Proc. 3rd Int. Conf. Expansive Soils, Haifa* **1**, 279–288. (NBRI R/Bou 466).

Donaldson, G. W. & Kleynhans, J. W. (1963). The progressive distortion of a house founded on heaving soil at Vereeniging. *Proc. 3rd Reg. Afr. Conf. Soil Mech., Salisbury* **1**, 149–152. (NBRI, RD 56).

Henrici, M. (1946). *Transpiration of South African plant associations*. Science Bulletin No. 247. Pretoria: Department of Agriculture and Forestry.

Henrici, M. (1947). *Transpiration of South African plant associations*. Science Bulletin No. 248. Pretoria: Department of Agriculture and Forestry.

Holland, J. E. (1982). *Economical housing foundation design on expansive soils*. Paper presented in South Africa under the sponsorship of the Natal Building Society.

Jacobs, M. R. (1955). *Growth habits of the eucalyptus.* Report of Australian Forestry School, pp. 141–143. Commonwealth of Australia.

Jennings, J. E., Brink, A. B. A. & Williams, A. A. B. (1973). Revised guide to soil profiling for civil engineering purposes in Southern Africa. *Trans. S. Afr. Instn Civ. Engrs* **15**, No. 1, 3–12. (NBRI R/Bou 455).

Mitchell, P. W. & Ruderics, A. J. (1982). *Designing house footings for the effects of trees.* Address to the Civil and Structural Group, South Australia Division of the Institution of Engineers.

Pile, K. C. (1981). Urban tree planting and the risk of pavement damage. *Proc. 1st Nat. Local Govt Engng Conf., Adelaide,* 196–199.

Pidgeon, J. T. & Tromp, B. E. (1983). Stiffened raft foundations for houses at the Loraine gold mine in the Orange Free State. *Civ. Engng Contractor,* Johannesburg. To be published.

Wiebe, H. H., Campbell, G. S., Gardner, W. H., Rawlins, S. L., Cary, J. W. & Brown, R. W. (1971). Measurement of plant and soil water status. Utah unsaturated soils. *Proc. Symp. A.S.T.M., Pennsylvania.* (Agricultural Experiment Station Bulletin 484).

Williams, A. A. B. (1980). Severe heaving of a block of flats near Kimberley. *Proc. 7th Reg. Afr. Conf. Soil Mech., Accra* **1**, 301–309.

Williams, A. A. B. & Simons, N. E. (1963). The heaving of roads on expansive clays with suggested remedies based on field moisture conditions. *Proc. 3rd Reg. Afr. Conf. Soil Mech., Salisbury* **1**, 47–51. (NITRR RR 49).

Wolpert, L. (1952). The direct measurement of the natural moisture content of unsaturated soils. NBRI Bulletin No. 9, pp. 47–73. Pretoria: Council for Scientific and Industrial Research.

APPENDIX 1. PROCEDURE FOR THE IN SITU DETERMINATION OF SUCTION

It has been stated that the in situ measurement of suction is preferable to laboratory determination. Although unstressed suction measurements are a convenient standard against which to compare the moisture content/suction relationships of different soil types, the effective in situ suction may differ significantly from the unstressed value. The unstressed values for remoulded and undisturbed samples can also only be expected to correspond closely in self mulching soils which are essentially remoulded in situ. Field measurement automatically accounts for temperature variations in the profile.

In recent years considerable attention has been devoted to devising a satisfactory method of installing the commercially available thermocouple psychrometer in order to undertake long-term monitoring of the variation of soil suction. Initial attempts using the fixed ceramic cup type were only partially successful, few surviving for more than 6 months. The problems encountered included contamination and corrosion of the thermocouple by salts drawn through or from the ceramic cup, following calibration using salt solutions, and the breakdown of insulation on the lead wires.

The introduction of the Wescor PST 55 and PCT 55 psychrometers coincide with a reassessment of the installation procedures. The PST 55 uses a special Dutch weave stainless steel screen to protect the thermocouple

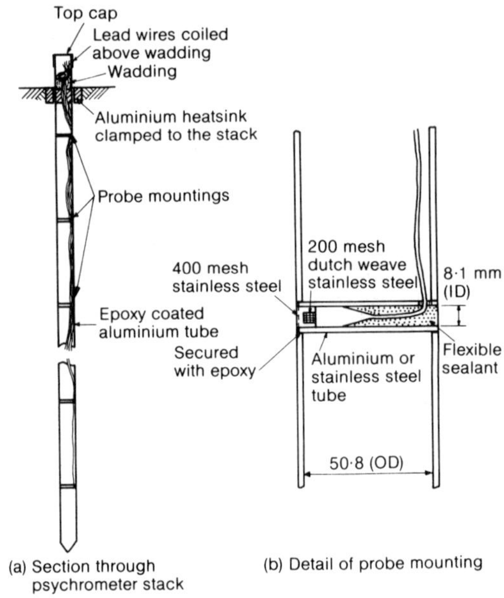

Fig. 8. Thermocouple psychrometer stack for in situ monitoring of soil suction

and this can be removed following calibration to clean the thermocouple. A 200 mesh is used for the screen and since this mesh size is permeable to water, problems could be encountered if free water containing salts inundated the psychrometer. To protect the probe from this potential problem, it is enclosed in a small aluminium tube with internal diameter equal to that of the vinyl boot of the probe. The closed end of the steel screen is located approximately 2 mm from the end of the tube and 400 mesh stainless steel is used to cover the end of the tube (see Fig. 8). A notch is provided in the opposite end of the tube with a diameter approximately double that of the lead wire and with carefully smoothed edges. Strain gauge covering putty is used to seal the rear end of the tube and to provide additional protection for the lead wires as they pass through the notch. The small tube is then fitted across the diameter of a standard aluminium irrigation tube (OD 50·8 mm) with the lead wire passing up the inside of the tube and with the ends of the tube secured with epoxy in exact size holes provided in the irrigation tube (care must be taken to ensure that the epoxy does not smear across the 400 mesh). The mesh is recessed from the face of the irrigation tube by approximately 0·5 mm.

A number of psychrometers may be fixed in this fashion in one tube creating a psychrometer stack. The stack is inserted in the ground after making a hole with a diameter equal to that of the stack using a thin walled sample tube with OD 50·8 mm; stress relief of the soil ensures a tight fit. In highly expansive soils, insertion should coincide with the driest season and soil condition in order to ensure that the soil does not shrink away from the tube. For additional protection the stack is sealed at the bottom by a welded aluminium cone and closed at the top with an aluminium cap, the psychrometer wires being retained at the top of the tube by wadding.

Calcrete in the Onderstepoort profile caused extensive oxidization of aluminium access tubes used for moisture content measurement with a neutron moisture probe. In some places holes had formed in the tubes allowing ingress of free water. To prevent this corrosion happening in the psychrometer stacks, the tubes were painted with an etch primer and epoxy paint.

The thermocouple psychrometer uses a chromel–constantan thermocouple mounted in a small probe behind a protective screen. Water vapour diffuses through the screen until the relative humidity in the cavity is in equilibrium with the water potential in the surrounding soil. A current is then passed through the thermocouple in one direction (exploiting the Peltier effect) to cool it below the dew point and water condenses on the thermocouple. After cooling is discontinued the psychrometric microvoltmeter can be used in two modes, either to measure the dew point temperature depression (hygrometric method) or the wet bulb depression (psychrometric method) which are functions of water potential. The dew point method has been used almost exclusively by the National Building Research Institute because it provides more reproducible results. In this method the current is automatically maintained in such a way as to prevent the moisture which has condensed on the junction from evaporating, or further moisture from condensing on it. The reading obtained on the instrument is proportional to the relative humidity and hence the soil water potential. Modern developments in the electronic circuitry have alleviated the problems of temperature compensation, although reasonable care is required.

The influence of vegetation on moisture and volume changes

I. RAVINA*

Soil moisture and volume changes in bare field soils and in the presence of vegetation are evaluated. In bare fields below the uppermost soil layer, which may dry up relatively fast, moisture may remain practically unchanged for a very long time if shrinkage cracks do not form. Deep-rooting plants can extract water from gradually increasing depths of up to 2–3 m, drying the soil to no lower than wilting moisture content, in general above the shrinkage limit. Accelerated drying rates under hot and dry weather conditions cause shrinkage crack formation in highly expansive soils. Crack formation is enhanced in cropped soils. The majority of shrinkage cracks extend to depths less than 50 cm, although depths of up to 135 cm are reported. Moisture and volume changes are observed to depths of 1·5–2·0 m in cracked soils. In the presence of deep-rooting crops the time required to reach maximum cracking volume is significantly shorter and the cracks are deeper and wider.

L'article évalue l'humidité du sol et les modifications volumétriques avec et sans végétation. Dans des champs sans végétation, au-dessous de la couche supérieure du sol, qui peut se dessécher relativement vite, l'humidité peut rester pratiquement inchangée pendant une période très prolongée, à condition que des fissures de retrait ne se forment pas. Des plantes qui s'enracinent profondément peuvent extraire de l'eau à des profondeurs qui s'accroissent progressivement jusqu'à 2–3 m, en desséchant le sol à une valeur qui n'est pas plus basse que celle de la teneur en eau accompagnant la flétrissure et qui est en général au-dessus de la limite de retrait. Des vitesses de dessèchement dans des conditions climatiques chaudes et sèches produisent des fissures dans des sols très expansibles. La formation des fissures est rehaussée dans des sols qui portent des récoltes. La plupart des fissures de retrait atteignent des profondeurs de moins de 50 cm, bien que des profondeurs de jusqu'à 135 cm aient été constatées. Des changements d'humidité et de volume sont observés jusqu'à des profondeurs de 1·5–2·0 m dans des sols fissurés. Avec des récoltes à racines profondes le temps nécessaire pour atteindre le volume maximal de fissuration est significativement plus court et les fissures sont plus profondes et plus larges.

INTRODUCTION

Volume changes in soil caused by changes in water content and in internal stresses affected by the water are of great interest in soil and foundation engineering as well as in agricultural engineering, soil–plant–water relations and pedology. Consequently, the amount of research carried out on this subject is very large and covers many aspects of the problem.

Like most physical processes in the soil, volume changes are greatly influenced by the amount and nature of clay present. Large volume changes are associated with high contents of smectite (montmorillonite). When such clay soils are dried, shrinking and cracking occur. Subsequent rewetting of the soil causes swelling, as pore water is adsorbed. Soils containing smectite show almost reversible swelling and shrinking on rewetting and redrying. Thus, the terms high swelling or expansive are applied to such soils.

A considerable amount of the knowledge currently available regarding the behaviour of expansive soils is a result of elaborate laboratory tests. Consequently, several mechanisms and models have been suggested to account for volume change and swelling pressure of clays as their water content is changed. Some examples of these mechanisms and models are interlamellar volume changes of an osmotic nature due to repulsion of adjacent diffuse double layers, differential strains between adjacent particles, rotational and translational displacements during wetting of randomly packed clay domains involving capillary cohesion and other sources for the driving forces beside double-layer repulsion (Yong & Warkentin, 1975; Quirk, 1978). The relative contributions of the various mechanisms to volume changes in the soil are as yet unresolved, and are beyond the scope of the present discussion. While the same mechanisms may be active, direct extrapolation of laboratory data collected on clay systems to the behaviour of field soils has met with only limited success and engineers continue to use empirical information.

Although conceptually the potential volume change of soil is considered to be an intrinsic property of a soil depending only on mineralogy, adsorbed cation species and concentration of the soil solution, in practice swelling and shrinking depend on many other factors. The magnitude of volume changes is governed by factors such as soil structure (total porosity and pore size distribution), initial and boundary conditions (initial water content, overburden and confining pressures), presence of cementing agents such as organic matter, and time and rate of water withdrawal.

* Technion-Israel Institute of Technology, Haifa, Israel

The property of a clay soil profile to shrink and swell under natural conditons has been well documented with case examples throughout the world, defining the important practical aspects of these processes. Volume change of a soil in situ will in general differ from that of a soil sample. The reason is failure to obtain exact duplication of field conditions in the laboratory. The environmental conditions of a particular area in which expansive soils are located play an important role in the behaviour of such soils. The frequency and intensity of rainfall or irrigation, the rate of evaporation, the depths of the groundwater table and the expansive soil layer and the non-uniformity of the soil are parameters in the eventual volume change. The nature and density of vegetation will directly influence the moisture changes in the soil and may also have an indirect influence on swelling and shrinking by affecting some of the related factors. The aim of this discussion is to review the processes of moisture depletion in the field and the influence of the presence of vegetation on the nature of the moisture changes and the associated volume changes and especially shrinkage cracking.

VOLUME AND MOISTURE CHANGES IN THE FIELD

When the soil in the field is wet, as it is at the end of a rainy period or irrigation, it will begin to lose water by evaporation. Evaporative losses from bare field soils occur at the surface, the evaporation process proceeding differently under various soil and environmental conditions (Hillel, 1980). Thus, for example, in the presence of a shallow groundwater table continual flow of water to the evaporative soil surface can occur without materially changing the soil moisture. However, more commonly, the loss of water at the surface and the resulting upward flow of water in the profile will cause the soil to dry; thus increase in suction and subsequently soil shrinkage will occur. However, once the top soil layer becomes dry and if shrinkage cracks do not develop, further drying is reduced considerably due to the low hydraulic conductivity of the drying soil, until finally water losses will be controlled by vapour diffusion, generally several orders of magnitude slower than water flow. The moisture content below the dry top layer will remain practically unchanged. A top soil layer of coarse crumbs having large pores and which dries up quickly can be most effective in conserving the moisture of the underlying fine soil. A mulch of plant residues, sufficiently thick, or of gravel, may also be helpful in suppressing evaporative losses from the soil.

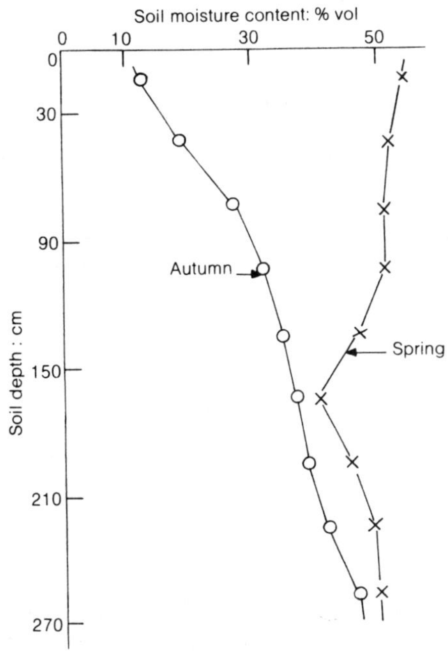

Fig. 1. Moisture content change in a grumusolic soil profile

Expansive soils with an aggregated top layer may occur naturally as a result of cyclic wetting and drying or freezing and thawing, or can be formed by proper cultivation. The stability of the crumbs is due to a sheath of oriented clay platelets on the sheared surfaces of the aggregates, which forms a plane of weakness. A similar sheath is formed on the walls of shrinkage cracks, so on redrying after each wetting the cracks will appear in the same places if the soil has not been otherwise disturbed. The swelling and shrinking characteristics of a clay soil with a crumb structure were investigated by Popescu (1980). The volume changes of an aggregated soil may be different from those of a structureless expansive soil because of differences in porosity and pore size distribution. A well aggregated soil with an appreciable non-capillary pore volume may show less than normal shrinkage for a wide range of moisture depletion. The predominance of smaller pores may enhance shrinkage by increasing capillary cohesion. They may also increase swelling due to the presence of entrapped air, while also diminishing the accomodation of microscale expansion into larger pores.

The cropping system can materially influence the size distribution of soil aggregates and soil porosity (Skidmore, Carstenson & Banbury, 1975; Harris, Chesters & Allen, 1966). Thus, it was found that a sod soil had greater porosity and less-dense aggregates than similar cultivated soils—cultivation resulting in a loss of pore space and a corresponding increase of bulk density. The soil aggregation status usually deteriorates rapidly if the soil is repeatedly cropped with annuals that supply little organic matter to the soil, require extensive cultivation and provide minimal cover. However, tilling the soil at its most favourable moisture content may improve the soil structure. Tillage may also prevent the development of shrinkage cracks or at least destroy them when they begin to form and thus diminish evaporative losses. The expansion and contraction behaviour of an expansive soil in the field with special attention to the cultivated layer was studied by Fox (1964a).

When shrinkage cracks develop during the process of soil drying, usually when moisture depletion rate is high, soil moisture in the profile tends to change to greater depths. The average moisture changes between spring and autumn in a cracking clay soil in Israel, as measured with a neutron probe over a period of five years, is shown in Fig. 1. Soil movements resulting from seasonal moisture changes were reported by Aitchison & Holmes (1953). They found that the amplitude of movement decreased with depth and that there was an increasing time lag with movement at depth compared to that at the surface. The swell at depth 1·2 m was 15% of that at the surface. Shrinkage was found to be slower than swelling and the soil at 1·2 m did not start shrinking until 3 months after the start of the dry season. Similar results were reported by Yaalon & Kalmar (1972) for grumusolic soil in Israel. A heave of 54 mm was measured at the surface, whereas at depths of 50 cm and 100 cm the soil moved 18 mm and 15 mm, respectively, after rewetting. The daily shrinkage during the dry season was 0·3 mm. A short diurnal heave and shrink effect was registered, apparently caused by temperature dependent volume changes of entrapped air.

When the soil surface is covered with vegetation the transfer of soil water to the atmosphere occurs mainly from plant canopies by a process called transpiration which takes place from the stomata of the plant leaves. Transpiration is caused by a vapour pressure gradient between the normally water-saturated leaf cells and the generally quite dry atmosphere. Soil water is extracted by the roots which permeate the soil and is transmitted through the stems to the continuously transpiring leaves. Thus, in the presence of vegetation the soil may dry more rapidly and to greater depths than a bare soil. Hence, vegetation can increase shrinkage and cracking. In a study of cracking of an expansive soil in Israel it was found that a soil inside a citrus orchard required only 42 days to reach the same cracking volume that developed in a similar bare soil during 160 days (Kalmar, 1979).

Like evaporation, transpiration is greatly dependent on meteorological conditions. When the soil is wet the transpiration rate will be determined mainly by the evaporation demand of the atmosphere. As long as the roots of the plant can extract water from the soil at a rate the canopy loses by transpiration, the plant will remain fully hydrated and maintain its normal activity. However, when the rate of water uptake by the roots falls below the rate of transpiration the plant will begin to lose water and if this situation persists it will cause the plant to wilt. As soil moisture is diminished and soil suction increases, the hydraulic conductivity may become too low to transmit soil water to the roots at the desired rate and/or the soil suction may exceed that of the root system and the soil will reach the wilting point moisture content. Although the soil suction at which plants wilt may not be the same under different conditions, the water retention characteristics of the soil enable us to define approximately the wilting point moisture content as the soil moisture at a water potential of −15 bars. At such a moisture content range, clay soils still retain sufficient water to cover all soil particles with several molecular layers of water. The shrinkage limit for most clay soils occurs at suction values in excess of 15 bars, thus in the field the full range of volume changes is generally not experienced. Grossman, Brasher & Franzmeier (1968) proposed a technique to evaluate the coefficient of linear extensibility (COLE) which is equal to $(lm - ld)/lm$, where lm is the length of the moist specimen and ld is its length when oven dry. Although the values obtained from such tests may be useful for identifying expansive soils they cannot predict volume changes of field soils since oven drying far exceeds the conditions to which such soils are exposed. Using the −15 bars moisture content instead of the oven dry and adjusting the confining pressure to values of overburden experienced in the field might improve the predictability value of such tests.

The amount and rate of water uptake depends considerably on plant properties. Most plants control their water balance mainly by developing extensive root systems while preserving an optimum ratio of roots to shoots. Thus, different plants have root systems varying in their density in the soil profile and the depth to which they extend. Water uptake from different depths in a soil of an initial uniform moisture content is not uniform. The growing roots of a young plant or an annual crop can advance continuously into moist regions of the soil and thus dry progressively larger soil volumes and deeper layers. Rose & Stern (1967) showed that during the early stages of growth of a cotton crop in the field, all water withdrawal was from the top 30 cm and, although later during the season water extraction extended to deeper layers, practically all the seasonal uptake was from the top 100 cm.

Although a mature root system occupies a more or less constant soil volume of fixed depth, the spatial distribution of the active absorbing roots can change markedly with time. Field observations (Ogata, Richards & Gardner, 1960; Taylor & Klepper, 1975) show that water is first removed from the top layer and with time from progressively deeper soil layers.

The bulk of the root system of most trees is within 100 cm of the ground surface. The majority of the active absorbing roots are found in the upper 30 cm of the forest soil, and the extent of lateral roots depends on competition between individual trees. In open-growing trees it is common to find lateral roots extending out two to three times beyond the radius of the canopy, although the majority of the active roots may lie within the area circumscribed by the periphery of the canopy (Kramer & Kozlowski, 1960). Otherwise, it was found that trees may affect the ground within a circle of radius about 1–1·5 times their height. Large structural settlements have been directly attributed to soil moisture withdrawal by vegetation adjacent to foundations. Ward (1947) and Skempton (1954) have suggested that trees should not be located closer to shallow foundations than their maximum height.

Bozozuk & Burn (1960) studied the vertical ground movement as a function of soil depth and distance away from a large (16·5 m high) elm tree. The vertical movement they found at depth 30 cm was as high as 52 mm at a distance of 1·5 m from the tree. Within a horizontal distance of 6 m from the tree the movements near the soil surface were still quite large, but they diminished rapidly with distance until at 12 m they were similar to the vertical movement found in a neighbouring grass plot (~12 mm). The soil movements also diminished with depth and at a depth of 375 cm the vertical movement was about 12 mm at both 1·5 m and 6 m away from the tree. Most turfs will have their root system in the upper 20 cm soil layer; however, weed grass may have deep rooting extending to even 1·5 m depth.

Moisture changes in the field are relatively limited and are confined to the ground upper layer of no more than 3 m deep. They are generally not uniform and thus cause uneven soil movement. Field soils are also rarely uniform in the vertical or horizontal dimension, nor is the distribution of water within them. Engineering problems associated with volume changes arise not only from overall change but also from differential shrinkage and swelling which are associated with such heterogeneities. Shrinkage and swelling are rate processes and depend on time. In the field, even in the absence of vegetation and cracks, swelling will seldom be completed and the maximum swelling pressure (as found in laboratory tests) will generally not be achieved.

Frequent rains or irrigation may reduce significantly the range of moisture changes both in time and in depth. Studies of volume changes of clods from irrigated and dry farming fields of a grumusolic* soil in Israel showed that irrigation may change the swelling–shrinking behaviour of the soil. The swelling potential of the dry farming soil profile was considerably greater than that of the irrigated field soil.

SHRINKAGE AND CRACKING

One-dimensional vertical shrinkage might occur in a plastic soil at a depth where the overburden consolidates the clay or when the soil is very slowly dried, but the more general case is approximately three-dimensional shrinkage accompanied by cracking. The dark clay soils known as grumusols or vertisols† are especially prone to cracking. They occur in regions with a wet–dry type climate, and although a dry season is a necessary feature, its duration may vary from throughout the year in the arid areas to only a few weeks in the wet regions. About 25 million hectares of such soils are found distributed on every continent (Dudal, 1965).

The high content of the highly expanding smectite clay (>30% clay in all horizons to a depth of at least 1 m) is of primary concern in the management of such soils. Engineering problems resulting from movement and distortion of buildings, highways, pipelines and utility lines as well as agronomic problems, such as possible disruption of root systems and increased evaporative water losses, are associated with vertisols. The formation of shrinkage cracks makes it nearly impossible to furrow irrigate such soils once they have become dry. On the other hand, cracking of agricultural field soils may be beneficial in promoting aeration of plant roots. Cracks may allow for rapid and deep percolation of water into the soil, thus becoming capable of rapidly transmitting water to considerable depths (Blake, Schlichting & Zimmermann, 1973). Ritchie, Kissel & Burnett (1972) have shown that wet field clay soils, where water movement may occur in invisible cracks and slickensides, have hydraulic conductivities 25 times

---

* Highly plastic clay.

† Cracking clay soils containing more than 30% smectites in all layers to a depth of at least 1 m, showing slickensides and parallelepiped structural aggregates. (US Department of Agriculture, 1974; see also Mitchell, 1976).

greater than the same soil which has been repacked in the laboratory. Cracking influences soil development and soil characteristics and is associated with gilgai micro-relief formation (Harris, 1964), churning or mixing of the upper part of the soil profile and the formation of slickensides and parallelepiped structural aggregates.

When a bare field of such swelling soils is allowed to dry on top, in due time a system of surface cracks in an irregular polygonal pattern will appear, characterized by an average spacing which varies from less than 1 m to more than 4 m in different soils and under different conditions. Once a crack begins, preferential drying and shrinkage inside it can cause the crack to gradually deepen and widen. Cracks assumed to develop in a bilateral shape, the base of which is the width of the crack on top, may be up to several centimetres, and the depth, as found by vertical measurement, can extend to 80 cm or more. Moisture changes in a cracking field are found to depths of 2 m. Micromorphological studies of grumusolic soils show at larger depths slickensides in a 45° orientation. Neither the changes in cracking characteristics of different soils under different conditions nor their deepening and widening nature have been investigated in detail. It would be useful to achieve a better understanding of the processes of cracking per se.

Not many works are found in the literature dealing with the influence of factors such as clay content, moisture changes or vegetation on soil cracking, and most of the information is based on general qualitative observations. The total crack volume is determined by soil moisture content and that more cracks develop when the initial moisture content is higher.

Textural variations affect the cracking pattern (White, 1972). The number of cracks per unit area depends on the mineralogy and the nature of the adsorbed cations. The moisture content of a soil at which cracking occurs was found to increase with increasing clay content. A greater degree of cracking was observed on cultivated areas than on bare fallow areas (De Vos & Virgo, 1969). In cultivated soils cracks were small and their maximum size was not reached until the soil had dried to the permanent wilting moisture content. It is suggested that cracks form where the cohesion of the soil is lowest (Yong & Warkentin, 1975). In clays with higher cohesion fewer and relatively larger cracks will form. When drying is not uniform cracks will form in the wetter soil. Observations indicate that planes of weakness leading to gross cracks occur in a soil where it dries out last. Thus, large cracks are observed in the border area between sod and bare soils. In a study of cracks in a road, Dagg & Russam (1968) found that cracks occur along a line between the zone where the moisture content fluctuates seasonally and the zone of more static conditions further under the pavement.

In a cultivated grumusolic field where crops are planted in rows it is found that a large continuous crack develops midway between plant rows, and across the rows smaller cracks occur extending from the main midway crack (Johnson, 1962). It was observed by Fox (1964b) that in no instance did these cracks intersect the plants. He also noticed that in the case of virgin grassland soils the major cracks surrounded the grass tussock and were circular in nature and again in no case did the smaller connecting cracks intersect the tussock. He concluded that the rooting system acts as a skeleton that holds the soil together, with cracking commencing at the junction of root systems. Studies have also shown that roots can significantly increase shear strength by increasing apparent cohesion of the soil (Endo & Tsuruta, 1969; Manbein, 1973).

The effect of natural vegetation on cracking was included in a thorough investigation of cracks in vertisols of the Sudan (Zein el Abedine & Robinson, 1971). The cracks in a site under grass and scanty trees were compared with those in a site of trees and scanty grass. It was found that although the volume of cracks in both sites was almost the same, the volume below 30 cm depth under trees was about 20% greater than that under grass. The effect of natural vegetation expressed itself also in the depth of the cracks: the area under grass had more than 60% of its cracks 16–45 cm deep, whereas under trees 75% of the cracks were 26–55 cm deep. The differences found in crack widths were limited to a narrow range of variability. However, in the site under grass the average crack width was 5·1 cm as compared with 4·8 cm under trees, indicating a greater surface dryness in the former. The soil under dominant grass cracked at an average spacing of 51 cm/1 m length, while the average spacing under trees was 39 cm (due to the mechanical binding effect of the shallow root system of the grass). In a field of wheat, which rapidly consumed the moisture from the top layer, deeper cracks (26–55 cm) were formed in approximately half the time shallower cracks (16–45 cm) developed on a fallow area. The volume and average depth of cracks was influenced by the amount and frequency of irrigation or precipitation. In a flood-irrigated cotton field the volume of cracks was 3–4 times smaller than in a non-irrigated soil. The maximum depth of cracks in the irrigated field did not exceed 65 cm, while in the non-irrigated area the cracks reached a depth of 135 cm.

The area of vertisol in Israel amounts to about 190 000 hectares, mostly arable and irrigated agricultural land. The long, hot, dry season causes extensive shrinkage and cracking. The shorter period (3–4 months) of rain results in swelling which was found to be as high as 70 mm (Kassiff, Livneh & Wieseman, 1969).

A study of the swelling and shrinking of grumusolic soil containing >60% clay, mainly smectite, also included measurements of depth, width, and spacing of the shrinkage cracks (Kalmar, 1979). Although moisture changes to a depth of 180 cm were observed (Fig. 1), the maximum measured depth of the cracks at the end of the dry season was 80 cm and 90% of the cracks had depths less than 40 cm. The first cracks were observed about 5 weeks after the end of the rainy season and their volume increased at a linear rate of about 1·2 m^3/hectare per day. The width of 90% of the cracks did not exceed 5·5 cm although some cracks were as wide as 10 cm. The maximum width was reached in about 12 weeks and remained the same until the end of the dry season, more than two months later. The crack spacing was also measured during the dry season. The first measurement showed uniform frequency distribution of spacings, varying from less than 20 cm to more than 1 m. When the numbers of cracks increased later in the season, the modal spacing was found to be less than 20 cm and 80% of the cracks were no more than 40 cm apart.

Micromorphological studies revealed oriented clay sheaths, 5–10 μm thick, on borders of the gross cracks and also of tiny fissures found at depths >70 cm, also in irrigated field soils. No such sheath was observed on cracks newly formed due to test procedures. Thus, field drying after wetting does tend to follow nearly the same cracking pattern.

The potential volume change of a non-irrigated profile which undergoes regular dry–wet cycles and cracking, as determined by the COLE test of undisturbed clods, is high throughout the 180 cm depth of the moisture changes. However, 15 years of irrigation considerably decreases the swelling potential of the soil especially below the top 30 cm.

CONCLUSION

Soil moisture changes in a bare field soil and in the presence of vegetation were evaluated. Except for the uppermost surface layer, field soils do not dry to less than the wilting point moisture content (approximately that of −15 bars). In a bare field and in the abscence of shrinkage cracks, the moisture content below the top layer may change only little over a long time. Deep-rooting plants may gradually extract soil water to depths of up to 3 m. Uneven soil movements associated with non-uniform moisture changes and soil heterogeneities are responsible for much of the damage to shallow foundations, structures and pavements. Vegetation will cause the soil to dry much more rapidly, thus enhancing the development of shrinkage cracks. Shrinkage cracks are common in areas of lithosol type soils, and are associated with many engineering problems. Measurements show that the majority of the cracks have depths less than 50 cm although depths of up to 135 cm are reported in non-irrigated field soils. In the presence of deep-rooting crops the time required to reach the maximum cracking volume was 4 times shorter than in a bare field. Frequent irrigation or rain may reduce the depth and range of moisture changes, and the associated volume changes, to a minimum. Cyclic swelling and shrinkage are essential for a favourable soil structure of grumusolic soils.

REFERENCES

Aitchison, G. D. & Holmes, J. M. (1953). Aspects of swelling in the soil profile. *Aust. J. Appl. Sci.* **4**, 244–259.

Blake, G., Schlichting, E. & Zimmermann, U. (1973). Water recharge in a soil with shrinkage cracks. *Proc. Soil Sci. Soc. Am.* **37**, 669–672.

Bozozuk, M. & Burn, K. N. (1960). Vertical ground movement near elm trees. *Géotechnique* **10**, No. 1, 19–32.

Dagg, M. & Russam, K. (1968). The relation between soil shrinkage and the development of surface cracks in an experimental road in Kenya. *Afr. Soils* **13**, 5–25.

De Vos, J. H. & Virgo, K. J. (1969). Soil structure in vertisols of the Blue Nile clay plains, Sudan. *J. Soil Sci.* **20**, 189–206.

Dudal, R. (1965). *Dark clay soils of tropical and sub-tropical regions.* Agricultural Development Paper No. 83, Rome: FAO.

Endo, T. & Tsuruta, T. (1969). The effect of tree roots upon the shearing strength of soil. *A. Rep. Hokkaido Stn Govt Forest Exp. Stn* **18**, 168–179.

Fox, W. E. (1964a). A study of bulk density and water in a swelling soil. *Soil Sci.* **98**, 307–316.

Fox, W. E. (1964b). Cracking characteristics and field capacity in a swelling soil. *Soil Sci.* **98**, 413.

Grossman, R. B., Brasher, B. R. & Franzmeier, D. P. (1968). Linear extensibility as calculated from natural-clod bulk density measurements. *Proc. Soil Sci. Soc. Am.* **32**, 570–573.

Harris, S. A. (1964). Seasonal density changes in alluvial soils of northern Iraq. *8th Int. Congr. Soil Sci. Bucharest* **2**, 291–303.

Harris, R. F., Chesters, G. & Allen, O. N. (1966). Dynamics of soil aggregation. *Adv. Agron.* **18**, 107–169.

Hillel, D. (1980). *Application of soil physics.* New York: Academic.

Johnson, W. C. (1962). Controlled soil cracking as a possible means of moisture conservation on wheat lands in the southwestern Great Plains. *Agron. J.* **54**, 323–325.

Kassiff, G., Livneh, M. & Wieseman, G. *Pavements on expansive clays*. Jerusalem: Jerusalem Academic Press.

Kalmar, D. (1979). *The dynamics of a heavy clay soil type grumusol*. PhD thesis, The Hebrew University, Jerusalem.

Kramer, P. J. & Kozlowski, T. T. (1960). *Physiology of Trees*. New York: McGraw-Hill.

Manbein, T. (1973). *The influence of soil moisture suction, cyclic wetting and drying and plant roots on the shear strength of a cohesive soil*. PhD thesis, University of California, Berkeley.

Mitchell, J. K. (1976). Fundamentals of soil behaviour. New York: Wiley.

Ogata, G., Richards, L. A. & Gardner, W. R. (1960). Transpiration of alfalfa determined from soil water content changes. *Soil Sci.* **89**, 179–182.

Popescu, M. (1980). Behaviour of expansive soils with a crumb structure. *Proc. 4th Int. Conf. Expansive Soils Denver*, **1**, 158–171.

Quirk, J. P. (1978). Some physico-chemical aspects of soil structure stability—a review. In *Modifications of soil structure* (ed. W. W. Emerson *et al.*). New York: Wiley.

Ritchie, J. T., Kissel, D. E. & Burnett, E. (1972). Water movement in undisturbed swelling clay soil. *Proc. Soil Sci. Soc. Am.* **36**, 874–879.

Rose, E. W. & Stern, W. R. (1967). Determination of withdrawal of water from soil by crop roots as a function of depth and time. *Aust. J. Soil Res.* **5**, 11–19.

Skempton, A. W. (1954). A foundation failure due to clay shrinkage caused by poplar trees. *Proc. Instn Civ. Engrs* Part 1, **3**, Jan., 66–83.

Skidmore, E. L., Carstenson, W. A. & Banbury, E. E. (1975). Soil changes resulting from cropping. *Proc. Soil Sci. Soc. Am.* **39**, 964–967.

Taylor, H. M. & Klepper, B. (1975). Water uptake by cotton root systems: an examination of assumptions in the single root model. *Soil Sci.* **120**, 57–67.

US Department of Agriculture (1974). *Soil taxonomy—a basic system of soil surveys*. Handbook No. 436. Soil Conservation Service. Washington DC: US Department of Agriculture.

Ward, W. H. (1947). The effects of fast-growing trees and shrub on shallow foundations. *J. Inst. Landsc. Archit.* **11**, 7–16.

White, E. N. (1972). Soil-dessication features in South Dakota depressions. *J. Geol.* **80**, 106–111.

Yaalon, D. H. & Kalmar, D. (1972). Vertical movement in a undisturbed soil: continuous measurement of swelling and shrinkage with a sensitive apparatus. *Geoderma* **8**, 231–240.

Yong, R. N. & Warkentin, B. P. (1975). *Soil properties and behaviour*. Amsterdam: Elsevier.

Zein el Abedine, A. & Robinson, G. H. (1971). A study on cracking in some vertisols of the Sudan. *Geoderma* **5**, 229–241.

The influence of vegetation on the swelling and shrinking of clays in the United States of America

W. G. HOLTZ*

Soils subject to swelling and shrinking can be found in most of the states in the continental United States of America, extending from the east coast to the west coast and from the north to the south. The cost of damage to structures placed on these soils is said to exceed the combined cost of all damage caused by earthquakes, tornadoes, hurricanes and floods. A significant portion of volume changes caused by shrinking and swelling can be attributed to the desiccation effects of vegetation. In the more humid areas of the USA, the greater portion of shrinkage is caused by transpiration from vegetation. In the arid and semi-arid parts of the country, evaporation is a major factor but transpiration losses can be very significant, often adversely tipping the damage scale. The effects of vegetation on soil moisture changes are very complex. These involve trees, shrubs and grasses, which may not only draw moisture from the soils, but may also cause changes in soil temperatures, soil chemistry, infiltration of water and soil moisture patterns. Problems associated with the removal of trees and other vegetation should be recognized with regard to damage corrections, new land developments and the establishment of new landscaping.

Des sols exposés au gonflage et au retrait se présentent dans la plupart des états composant les Etats-Unis d'Amérique. De tels sols sont à trouver sur une superficie s'étendant de la côte est à la côte ouest du continent, du nord jusqu'au sud. On dit que le coût des dégâts causés aux édifices bâtis sur de tels sols dépasse celui qui est le résultat de tous les dégâts produits par les tremblements de terre, les tornados, les ouragans et les inondations pris ensemble. Une proportion importante des changements volumétriques causés par le gonflage et le retrait peuvent s'attribuer aux effets dessicatifs de la végétation. Dans les régions plus humides des Etats-Unis la plupart du retrait est causée par la transpiration de la végétation. Dans les régions arides et demi-arides du pays l'évaporation représente un aspect d'importance primordiale, mais les pertes dues à la transpiration peuvent être considérables et sont souvent responsables du déséquilibre qui crée les dégâts. L'influence exercée par la végétation sur les changements dans l'humidité du sol sont très complexes. Les arbres, les buissons et les herbages peuvent non seulement tirer de l'humidité des sols, mais ils peuvent aussi produire des changements en ce qui concerne les temperatures et la chimie des sols, l'infiltration de l'eau et la distribution de l'humidité. Ceux qui sont chargés de corriger les causes éventuelles des dégâts ou d'ébaucher de nouveaux développements affectant les terrains devraient se rendre compte des problèmes entourant l'éloignement des arbres et d'autre végétation aussi bien que des problèmes qui surviennent lors de l'établissement d'un nouveau paysage.

INTRODUCTION

The influence of vegetation on the swelling and shrinking of soils is a well known topic but one on which there is little quantitative documentation. Jones & Holtz (1973) detailed the high cost of damage in the US that is caused by soils which swell and shrink. The annual loss estimate of $2255 million was considered by some engineers to be about half the actual value. In terms of 1982 dollars the loss figure now may be closer to $6000 million annually. The losses were estimated to exceed those caused by earthquakes, tornadoes, hurricanes and floods combined. Within the average American's lifetime, there will be a 14% reduction of the land by the above four disasters but over 20% of the land will be affected by the shrinking and swelling of clay soils. About 20% of this damage, or about £1200 million, can be traced to desiccation effects of vegetation.

Soils subject to shrinking and swelling are said to be found in nearly every state in the continental US. A 1972 survey of state highway departments by the Federal Highway Administration indicated that 36 states had expansive soils within their geographical jurisdiction, e.g. Virginia, the Great Lakes areas of Illinois and Wisconsin, South Dakota, Wyoming, Colorado, Utah, Nevada, Northern California, Georgia, Florida, Mississippi, Oklahoma, Texas, New Mexico, Arizona and Southern California. Canada and Mexico also have their soil volume-change problems.

The greatest portion of damage from expansive soils is found in the arid and semi-arid western parts of the US. This is attributed to two factors—the formation of montmorillonite minerals and the possibility of major changes in soil moisture during annual and longer-term, cyclic precipitation events. While desiccation from trees and other vegetation is important in these regions, it can become the major damage reason in much of

* Formerly US Bureau of Reclamation, now Woodward-Clyde Consultants Inc, Denver.

the eastern part of the country where there is greater precipitation and more humid climatic conditions and therefore, less evaporative soil-moisture losses.

SOIL TYPES

Expansive soils which are subject to volume changes on wetting and drying are considered to be soil or rock materials in which clay minerals, white alkali or pyrites constitute a significant fraction of the soil or rock. Fine clays having illite, kaolinite and montmorillonite minerals all have some potential for changes in volume when moisture changes take place. Sodium montmorillonite type clays have a potential for large volume changes and are considered to cause the worst disruption of structures placed on them. These clays include shales with montmorillonite minerals and soils derived from them.

Swelling and shrinking clays are normally identified in a semi-quantitative manner by Atterberg tests for plasticity index and liquid limit (Holtz & Gibbs, 1956; Seed, Woodward & Lundgren, 1962). Shrinkage limit tests which measure the amount of shrinkage from about the liquid-limit state to an oven-dry state are also helpful (Holtz & Gibbs, 1956). For more quantitative data for design purposes, consolidometer swell tests (ASTM, 1982) or suction type tests are generally used. Surficial geologic information is commonly used to define areas that are suspect.

VEGETATION EFFECTS

Desiccation of clay soils subject to significant shrinking on drying and swelling on rewetting is usually thought of in terms of trees and large shrubs with respect to their root systems and the transpiration of moisture through the leaf systems. However, there are several phenomena related to vegetation effects.

The first of these are trees which withdraw moisture from the soil to satisfy their moisture demands. As trees grow their moisture demands become greater, and thus damaging effects become apparent and increase with time. Large trees draw more moisture than small trees or shrubs. Influence areas also are affected by whether the root systems are shallow or deep. For instance, a troublesome clay layer overlaid by non-troublesome soil could generate a problem if deep rooted trees penetrated it, while shallow rooted trees and shrubs might not have an effect. The root system of trees extend out to their drip lines and it is here where moisture withdrawal could be the greatest. However, moisture can be drawn from outside this area as suction pressures develop and soil moistures tend to equalize. Under dry conditions roots will tend to extend to wherever they can find moisture.

The withdrawal of moisture by trees, shrubs and other forms of vegetation is usually seasonal, particularly with deciduous varieties and transpiration occurs mostly during the growing season. In such cases soil moisture replacement may occur during the dormant season, but such replacement may not be sufficient to totally replace the loss. This depends on, for example, the weather cycle, the geographic location (long or short growing seasons), climatic conditions, ground water conditions, soil permeabilities, location and drainage from nearby structures. In any case, soil moistures can vary seasonally and annually with resulting volumetric changes in the soil, which often contribute to nearby building damage. Evergreen varieties will withdraw moisture to a certain extent throughout the year.

Trees and bushes shade ground surfaces and thus reduce ambient soil temperatures, resulting in the lowering of evaporation rates. Vegetation builds up organic material in its vicinity and this acts both as a sponge to retain surface moisture and also as an insulating medium (Jones, 1979).

Lawns and other grasses affect soil moisture. They retard precipitation runoff as compared to bare ground, but also contribute to soil moisture depletion during dry periods. They contribute to organic buildup and encourage earthworm populations with resulting channels for increasing moisture penetration. With trees, bushes and other types of vegetation, the disintegration of roots provide channels for water penetration.

Alkali soils and pyrites exhibit volume change properties. The chemistry of soils affects not only volume change characteristics but also the type of vegetation that the soils can support. The US Geologic Survey (USGS) has conducted research on plant growth in the western US with respect to various soils, soil chemistry and osmotic stress for different plant communities. The USGS has produced numerous publications on the subject, e.g. Branson, Miller & McQueen (1970). These studies are mainly concerned with improving rangelands. While they are not necessarily related to shrinking and swelling of soils, they do relate to soil desiccation and have information on plant effects on soil moisture profiles and osmotic stresses for various types of soils and plants.

When vegetation is removed and moisture withdrawal ceases, soil moisture will tend to increase. The clay soils will then generally expand as the previously desiccated bulb tries to return to its natural condition. This is a particularly important consideration when new areas are being developed for housing, businesses, roads or other structures. Other factors which can affect clay soils in new developments are changes in precipitation runoff patterns, shade from buildings and the irrigation of lawns and other vegetation. Changes in ground water patterns caused by the construction of basements and deep foundations have also had noticeable effects on soil moisture and clay behaviour.

All factors which change soil moisture patterns can have an effect on the shrinking–swelling phenomenon at a specific location where clay soils are involved. They should be given careful consideration by the geotechnical engineer.

DESICCATION DAMAGE

Geotechnical engineers who are faced with the design of structures on shrink–swell clay soils, or the remedial repair of damaged structures on these soils, have made studies related to specific structures or troublesome areas and many deal with desiccation effects. A few of the latter are discussed herein as examples.

Holtz & Hart (1978) discussed recommended practices for landscaping and the planting of trees and shrubs at proper distances from houses. Such plantings were discussed from two standpoints in the semi-arid Colorado–Wyoming area: shrinkage from plant transpiration and swelling caused by irrigation, near the buildings. As a general rule for the Colorado–Wyoming area, it was recommended that grass, shrubs and flowers be kept a minimum of 5 ft from the building. To control desiccation all trees should be planted at least 15 ft from any building, except large water-loving trees which should be kept at least 20 ft away.

Barber (1956) reported on the shrinkage of the marine Tuxedo clay in the coastal plain near the District of Columbia. Here some buildings were damaged by the swelling of the clay and several other structures were cracked by the drying of the same clay when aided by the roots of trees.

Perpich, Lukas & Baker (1965) detailed damage caused by desiccation from trees in the glacial lake bottom and ground moraine soils of the Chicago, Illinois area and included some useful information on the transpiration rates of various types of trees, experimental soil suction values developed and the depth and lateral extent of various root systems. Fifteen cases of damage by desiccation were reviewed. Over 200 structures and foundations were investigated where desiccation from tree growth was a factor. It has been found that certain trees have roots that not only extend well beyond the branches, but also penetrate deeply into the soil. A root system may extend into a volume of soil as large, or even larger, than the space occupied by the above-ground part of a tree. It is also shown that certain broad-leaf deciduous trees have transpiration rates two or three times that of certain evergreen trees (Kramer & Kozlowski, 1960).

In the Chicago area, which has an average annual rainfall of about 34 in., transpiration rate becomes the most important factor of soil drying as compared with semi-arid areas in which both evaporation and transpiration are major factors. For comparison, the Denver area has a precipitation of about 13–15 in./year. In parts of western Colorado where the precipitation is only about 9 in./year, the pan evaporation rate is three times that amount. However, desiccation by transpiration of plants and trees can compound soil moisture losses.

In the Chicago area damage begins to occur when the soil moisture becomes less than the plastic limit. Given their climatic conditions, damage to buildings was caused by trees having a trunk diameter of over 10 in.; small trees and shrubs were not a problem. In no cases were evergreen trees involved. It was recommended that trees should not be placed closer to a structure than the tree height (Perpich *et al.*, 1965). This is generally a safe, conservative distance.

Remedial measures for desiccation shrinkage damage includes the removal of trees, which is probably the best approach. However, if owners do not wish to lose their trees, an alternative approach is to restore water to the soil and tree root system by, for example, irrigation through sand filled holes and perimeter trenches. Underpinning below the zone of influence can be done but this is costly and usually limited to large structures. When trees are removed, or irrigation applied, the soils will regain some of their natural moisture. There is then the potential for these desiccated soils to swell which creates the potential for swell uplift to nearby slabs and shallow footings in particular. This risk should be recognized.

St Hedwig's Church, Chicago, suffered the separation of one 150 foot high Gothic tower from the main building. After it was determined that the movement was caused by soil desiccation, the suggested underpinning was delayed until an experimental irrigation remedy was tried. After irrigating the tower foundation area the tower returned to its original position and the cracks closed.

The desiccation of clay soils with resulting structure damage has also been observed in the vicinity of Fond du Lac, Wisconsin. Here the soils are a part of the Valders drift and consist largely of the Glacio–lacustrine portion of the drift. The sediments are generally high in montmorillonite-type clay content. While the major damage to buildings has been due to desiccation from trees there are instances where dried soils have expanded on wetting.

While desiccation in areas of medium to high precipitation may be limited to large deciduous trees near structures, grasses, shrubs and both small and large evergreen and deciduous trees can cause damage to foundations and slabs by desiccation in arid and semi-arid parts of the country. Shrinking of as much as 2 in. under slabs has been commonly noted in the Denver area during dry periods. Such shrinkage starts at the edges of walks, kerbs, driveways and patio slabs and progresses inward, the extent depending on the drying period. When wet cycles return the clay soils expand. Thus, there is continual movement over periods of time. Cracking and offsetting of the shallow structures occur. Shallow foundations and floor slabs can also go through cyclic changes. Damaging desiccation takes place alone from evaporation of soil moisture, but additional transpiration of soil moisture from nearby trees and shrubs can increase structural damage or make the damage more objectionable.

Texas has the largest areas of shrinking and swelling soils in the US. This is not only because it is the largest state but also because of its geologic background and its climatic conditions. With its wide variation of precipitation, being high in the humid east and low in the dry west, different types of problems are encountered. Because of the widespread shrinking and swelling effects on highways, buildings and other structures, a major amount of research has been carried out on expansive soils at the Texas A & M University, College Station, the University of Texas, Austin, the Texas Highway Department, and others. Studies have included desiccation of the clay soils by trees and other vegetation.

Felt (1953) was one of the earliest to report on the influence of vegetation on soil volume changes. Shrinkage and swelling was reported on the highway in McLennon County, Texas. Both large trees and lush vegetation along the highway right-of-way contributed to the problems. Where trees, bushes and grasses had been removed prior to construction, lumps in the pavement began to appear when the abnormally dry soil under the pavement slab regained moisture and swelled. Another example was where heavy grass vegetation along a highway was responsible for excessive loss of moisture from clay under the edge of the slab. Shrinkage of the soil and attendant loss of support for the pavement resulted in considerable longitudinal cracking.

Rational methods have been developed for designing building slabs and foundations, highway and airport pavements and certain other structures placed on shrinking–swelling soils (Castleberry, 1974; Dobson, 1978; Lytton, 1982).

With this approach the foundations and pavements are designed to meet an acceptable level of risk. In addition to considering the structural characteristics of the structure, independent variables in the damage equations include numerical measures of drainage, rainfall, slope, soil properties, depth of active zone, age, shape and type of building, owner's care and number of trees, as well as their spread and their distances from the closest wall of the structure.

To study these variables and arrive at usable numerical measures, many buildings, other structures and sites were studied in locations at or near Houston, Beaumont, College Station, Waco, San Antonio and Amarillo, Texas. The effects of pre-construction and post-construction vegetation are well documented and are exhibited in terms of plots of damage severity index* as related to the percentage of vegetation at a particular location (Castleberry, 1974; Dobson, 1978). In some cases the damage rating is given with respect to average distances of trees from houses.

Castleberry (1974) gives an example where the severity index exhibited an increase with pre-construction vegetation, indicating that ground moisture conditions were still adjusting to the removal of large trees after a slab had been set. This is an important consideration for structures built in newly developed areas where trees and other vegetation had existed previously.

The Texas studies showed that the severity of damage increased with the percentage of vegetation and the closeness of trees. For instance, in the Waco model it was noted that the occurrence of three post-construction vegetation variables had their effects on soil moisture conditions and resulting severity of damage: percentage of lot covered by vegetation, number of trees on the lot and average distance of trees from the house. The number of trees had a lesser effect than the amount of vegetation coverage. The influence of small shrubs and brushes on the soil moisture conditions should not be overlooked. This is an important factor that has increased importance in semi-arid and arid environments.

Research studies were conducted at the University of New Mexico on the design of airport pavements on expansive soils (McKeen, 1981). Information is included on vertical soil movements associated with soil–water depletion at the Gallup, New Mexico and Dallas/Fort Worth airports. The effects of changes in soil suction on elevation changes are plotted. These suction changes are due

* The index rates damage severity from 0 to 4. 0 denotes no observable deformation in slab or veneer; 4 denotes deformation associated with slab cracks having vertical or horizontal separation implying loss of structural integrity .... Major damage to home requiring extensive repair or possible condemnation.

to normal soil–climate interaction and transpiration by native grasses. It is reasonable to expect that trees would transpire additional water and reach greater depths. Therefore, vertical movements could easily exceed the 0·2 ft shown, if trees were located nearby.

California has been noted for uplift and shrinkage damage and landslides caused by expanding and shrinking clay soils. Locations range from the Los Angeles area up through the Central Valley and into the upper Bay area. Contra Costa County is one of the areas where highway pavements and canal linings have suffered from expansion. In a personal communication Dr J. Mitchell, University of California, Berkeley, reported a case involving a slope movement caused by clay soil desiccation. This happened during the northern California drought, when normal watering of yards was restricted. Here a scarp developed on a hillside road which had the appearance of a normal slope failure. There was a group of pine trees at the bottom of the slope in the backyard of a house. Borings along the roadside indicated low water contents in this area and that shrinkage, made worse by the trees, was the problem.

When such drought cycles occur, trees and other vegetation attempt to survive by extending their root systems deeper and broader and they also develop higher soil suction pressures. Thus the effects on soil moisture changes are magnified. This increases the need to consider long range precipitation cycles as well as seasonal precipitation events.

CONCLUSIONS

Vegetation can have damaging effects on slabs, buildings, pavements and other structures founded on soils having shrinking and swelling potentials, by causing the depletion of soil moisture. Large, broad-leaf, deciduous trees located near the structure causes the greatest changes in moisture and the greatest resulting damage in both arid and humid areas. Small trees, bushes and grasses also create sufficient transpiration rates to affect soil moistures at shallower depths, particularly in arid and semi-arid areas. The effects of this type of vegetation must be given consideration in any rational evaluation. The total vegetation coverage of a site as well as the number, size, location and type of trees also affects soil moisture losses and resulting structure damage risks.

Vegetation has several effects on soil moisture. In addition to moisture removal by transpiration, the shading of the ground surface, buildup of organic material, retardation of precipitation runoff, and the formation of water channels by root disintegration all can cause changes in soil-moisture patterns.

In new development locations, the removal of trees and other vegetation can result in the swelling of troublesome clay soils as the soil moisture tends to return to its natural state and as additional new water is made available.

The rational and quantitative evaluation of vegetation effects on soil moisture, the resultant effects on the shrinking and swelling of certain clay soils, and the resultant damage to structures on these soils is very complex and not easy to formulate. More research is required in this area of geotechnical engineering.

REFERENCES

ASTM, (1982). One-dimensional expansion, shrinkage and uplift pressure of soil-lime mixtures. Designation D-3877-80. In *Book of Standards*, Part 19. Philadelphia: American Society for Testing and Materials.

Barber, E. W. (1956). Discussion of paper by Holtz & Gibbs, titled Engineering properties of expansive clays. *Trans. Am. Soc. Civ. Engrs* **121**, 669–673.

Branson, F. A., Miller, R. F. & McQueen, I. S. (1970). Plant communities and associated water factors on shale derived soils in northeastern Montana, US Geologic Survey, Denver, Colorado, *Ecology* **51**, No. 2.

Castleberry, J. P. (1974). *An engineering geology analysis of home foundations on expansive clays.* MS thesis, Texas A & M University.

Dobson, B. M. (1978). *The influence of stratigraphy on the behaviour of expansive soils.* MS thesis, Texas A & M University.

Felt, E. J. (1953). Influence of vegetation on soil moisture contents and resulting volume change. *Proc. 3rd Int. Conf. Soil Mech., Zurich* **1**, 24–27.

Holtz, W. G. & Gibbs, H. J. (1956). Engineering properties of expansive clays. *Trans. Am. Soc. Civ. Engrs* **121**, 641–663.

Holtz, W. G. & Hart, S. S. (1978). *Home construction on shrinking and swelling soils.* National Science Foundation, Washington DC. Denver: Colorado Geological Survey.

Jones, D. E. Jr & Holtz, W. G. (1973). Expansive soils—the hidden disaster. *Civil Engineering* August, 49–51.

Jones E. E. Jr (1979). The expansive soil problem. *Underground Space* **3**, No. 5, 221–226.

Kramer, P. & Kozlowski, T. (1960). *Physiology of trees.* New York: McGraw-Hill.

Lytton, R. L. (1982). *The design of foundations and pavements on expansive clays*, Texas A & M University. Unpublished.

McKeen, R. G. (1981). *Design of airport pavements for expansive soils*, Report No. DOT/FAA/RD-81/25. Washington DC: Systems Research and Development Service.

Seed, H. B., Woodward, R. J. & Lundgren, R. (1962). Prediction of swelling potential for compacted clays. *J. Soil Mech. Fdns Div. Am. Soc. Civ. Engrs* **88**, 53–87.

Discussion

R. WYLES

There have been two developments of relevance to the Paper.

Ketteman V. Hansel Properties Limited

This judgement,[1] although a decision of a deputy official referee, is of importance because it is the first construction case to interpret Pirelli.[2]

The facts briefly were that five properties were constructed in June 1975 and by mid-1976 these had shown signs of structural damage due to foundation defects (there were trees nearby and the soil was clay). The builder was immediately notified, but in the event no remedial work was done, and a writ was issued against the builder in 1980. On 25 June 1982, the architect and local authority were joined in the action as defendants. The builder had by that time ceased trading.

For the purpose of the Limitation Acts, the plaintiffs had to establish that their cause of action arose after 25 June 1976, i.e. for the purpose of Pirelli,[2] that the damage first occurred after that date.

Three of the plaintiffs succeeded against the architect. It was held that as notice of the damage was given to the builder in early August, the damage must have occurred only one week before then, i.e. after 25 June. The two remaining plaintiffs were unable to prove that damage first occurred after 25 June and their claim therefore failed.

There are two points to note about this decision. Firstly, the judge accepted—presumably because no technical evidence was adduced to the contrary—that if cracks appear consequent upon foundation failure damage occurs, i.e. the foundations fail, one week before the manifestation of the cracks. From a technical point of view, this decision appears to be suspect, and reinforces the argument that if a Pirelli defence is being pleaded, the defendant must provide detailed technical evidence to show when, in all probability, the damage must first have occurred.

Secondly, the claim against the local authority failed because at no stage was there any risk to health and safety of the occupants of the dwellings, even though it was accepted that underpinning at a cost of over £10 000 per property was necessary.

New draft building regulations

New draft Building Regulations have been circulated for preliminary comment. The main point to note is that they are drafted as simple functional requirements and that detailed instruction on how these requirements are to be complied with will be set out in approved documents. These will be approved by the Secretary of State, having been issued by bodies such as the British Standards Institution and the Agreement Board. It is relevant to note here that in the draft Approved Document for sites, it is stated that the recommendations in the 1980 code of practice (British Standards Institution, 1980) are to be followed when necessary. Compliance with this code will mean, therefore, that the builder will be deemed to have complied with the Building Regulations.

R. DRISCOLL

Since the Paper was written, new data on case histories 1 and 2 have been obtained. Fig. 7 in the Paper shows crack progression in Case 1; the latest readings of crack width are plotted in Fig. 1 and show the effect of tree pruning in September 1982. There is some uncertainty about the precise degree of pruning, but the foliage is thought to have been reduced in volume by 50–75%. The overall 1982 summer increase in crack width can be seen to have been significantly reduced in comparison with increases in 1980 and 1981. Furthermore, the August 1983 reading shows the crack to be continuing to close, whereas previous, corresponding summer readings have shown the crack

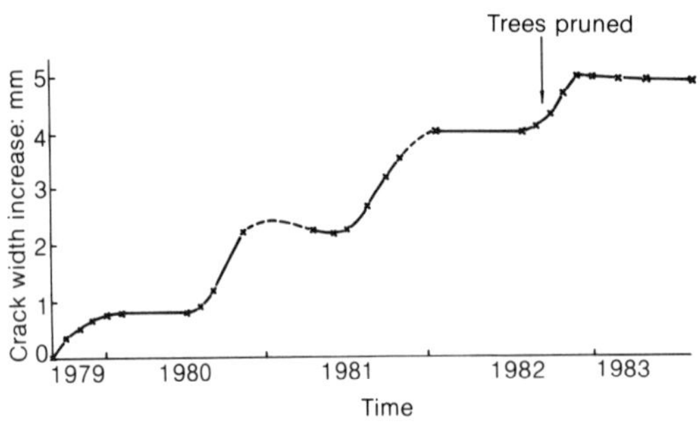

Fig. 1. Crack width progression in Case 1

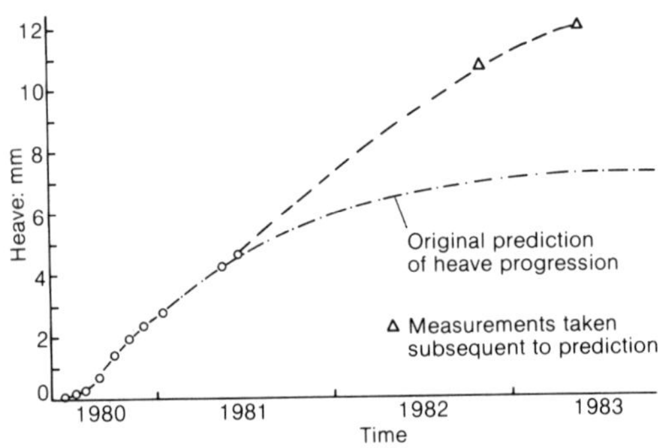

Fig. 2. Heave measurements at point X in Case 2

to be widening. Also, the summer of 1983 was comparatively dry, so that the trees, were it not for the pruning, could be expected to have had a greater effect than in the previous three summers. It would appear, therefore, that pruning has affected a cure in this case.

Figure 2 shows the latest level readings for Case 2. It had been predicted in the Paper that, after nine years of heaving, movements would cease in 1982 (see Fig. 2). However, levels taken in 1982 and 1983 show substantial additional heave, with no clear indication of imminent cessation.

A. D. M. PENMAN (geotechnical consultant)

I was investigating the problem of damage to houses on shrinkable clay caused by trees in the late 1940s and early 1950s, when I was working with Ward at the Building Research Station.

In those days it was accepted that trees were responsible for the damage. However, nowadays, we have to contend with sensational journalism, all too often typified by such eye-catching headlines as 'Scaffolding collapses, man killed, engineer to blame', and nonsense in our law. So, whereas in the old days when a house built on shrinkable clay cracked and a tree might be blamed, now the fault is laid on the engineer.

It was often found that householders had planted fast growing poplar trees in a row to form a screen from their neighbours. If severe structural damage occurred when the trees had grown to the height of the houses, this was clearly not the fault of the designer of the house foundations if they had proved satisfactory prior to the trees.

It seems a nonsense, therefore, that an engineer may be held responsible for damage that occurred 22 years after foundations he designed (or the design of which he approved) were built and an even greater nonsense for his liability to hinge on whether or not action is initiated within 6 years of damage first occurring, even if at that time it was not detected.

I see that many local authorities now engage private consultants to make the required checks on the designs of buildings submitted for Building Regulations approval. They say that they are driven to do this by economies and cut backs which have left them with insufficient staff. Is this perhaps a method whereby responsibility is transferred so that claims cannot be made against the authority if foundations fail?

E. J. SKILTON (John F. S. Pryke and Partners)

I have two points to make in regard to Miss Wyles' paper, both of which revolve around recent interpretations of the law.

My first point concerns the Acrecrest[3] case. Both the High Court and the Court of Appeal seemed to accept that neither architect nor building inspector should have been aware of the dangers of Dutch Elm disease and that they could not have reasonably foreseen the dangers of heave in consequence. The facts in this case were complex and involved damage to buildings constructed in about 1971 on a site where some trees were removed, some pollarded and others subsequently died. I would suggest that it is reasonable to expect all trees to die eventually, most through natural causes but some prematurely due to disease or accident. Indeed, the dangers of heave as a result of tree removal are well documented in the technical literature prior to 1970 and it would seem to be unwise therefore to rely on the wording of this part of the judgement as a defence against failure to consider such effects.

The present liability of local authorities and professionals in this field may have been increased by the publication of a code of practice for trees in relation to construction (British Standards Institution, 1980). Clause 8.1 warns of the possibility of buildings coming into close proximity to trees even as a result of self-sowing, whilst clause 8.4.2.2 states that where shrinkable clays are present and roots are found, *or are anticipated in the future* (my emphasis) then precautionary measures should be taken in the design of the building's foundations. Since this code represents a standard of good practice, it is essential that engineers engaged in the design of new foundations, or repairs to existing foundations, should take reasonable account of future root growth in their designs. Exactly where the delineation of reasonableness stands may be for the law to decide, but with the trend for courts to find in sympathy with the house owning individual, engineers can be forgiven for erring on the side of caution.

My second point concerns the Bluett[4] case and the thorny question of whether repairs are necessary. According to the Paper, damage was only slight; presumably this means cracks less than 5 mm in width (Building Research Establishment, 1981).

While accepting that there has in the past been a certain over-reaction by some engineers to the need for expensive repair works, there is an overriding and possibly conflicting duty of house owners in these circumstances and that is to mitigate their losses. If a property becomes blighted or unsaleable at the full market value as a result of cracking and foundation problems and, provided underpinning cures these problems and is a reasonable solution suited to the house owner, then this can be undertaken irrespective of the degree of damage, although in my experience crack widths of 2–3 mm can be considered as the average threshold of alarm.

Buildings that have suffered from foundation movements in the past, albeit slight, are considerably more 'at risk' in the future and insurers may refuse to grant subsidence or heave cover under these circumstances. In other words, it is market forces that will dictate what is a 'reasonable' course of action and not the BRE classification, the terminology of which does not gain general acceptance from the public.

My opinion is to some extent backed up by a judgement in a case[1] where the remedial works (underpinning) scheme was challenged. Estate agents' and insurers' views were taken into account in this case and, in addition, the judge indicated a requirement to 'put the plaintiffs in the position in which they would have been had the house been constructed properly'. Since this would have required compliance with the Building Regulations (Building Regulations, 1976) and, in particular, the often quoted regulation D3, I would submit that cracking of some 3–5 mm in walls (and it was more in the case quoted) is not a characteristic of the average domestic residence and is therefore unacceptable.

I. STATHAM and C. P. THOMAS (Geotechnical Engineering Ltd)

R. Wyles

The Paper and subsequent verbal discussion hold serious implications for foundation recommendations on shrinkable clays. It appears that current interpretation of existing practice notes, for example BS 5837 (British Standards Institution, 1980) is that foundations must take account of problems associated with trees, even though no trees may be present at the date of, or prior to, investigation of a site. Taken to its logical conclusion, this implies that all foundations need to be designed for a mature tree of high water demand being planted (or allowed to grow) very close to the structure. In other words, the worst possible case of desiccation and settlement must be anticipated to avoid the possibility of being found negligent in one's recommendations. If this is so, all foundations on shrinkable clays will need to be designed to accommodate potential movements in the zone of maximum potential desiccation, a variable and difficult parameter to establish if no desiccation exists at the date of investigation, irrespective of actual ground conditions. The implications for foundation costs and client incredulity are only too clear.

On the one hand, professional engineers are legally expected to be increasingly aware of potential problems associated with trees and to suffer the consequences if they are not, while on the other we are urged by Biddle and Driscoll to relax our recommendations. It is essential that this discrepancy in current published work and interpretation of current practice is resolved as a matter of urgency. Further, if we accept Driscoll's argument that to preclude 75% of cases of damage caused by tree roots is more realistic than total preclusion, we can probably all look forward to even higher professional indemnity premiums in the future.

R. Driscoll

The Paper presents a useful summary of the development of awareness of the problems relating to trees on shrinkable clays. However, his guidance upon the detection of desiccation in clay soils, in our opinion, is not really adequate to be used in practice, and also contains some inconsistencies.

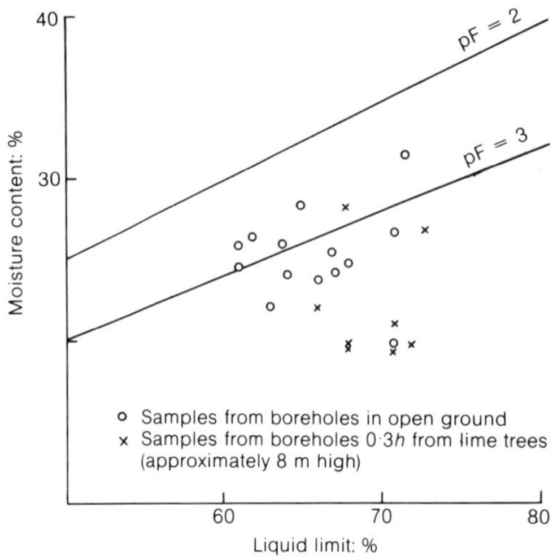

Fig. 3. Moisture content plotted against liquid limit for samples of Lower Lias clay near to and remote from lime trees

Driscoll is right to draw attention to the possibility of being misled by swelling pressure test results, since overconsolidated clays will often show swelling pressures greater than effective vertical overburden pressures, simply because K_0 is more than one. His approximate expression for calculating the theoretical swelling pressure of an undesiccated clay is based on the (unstated) assumptions that the porewater pressure coefficients A and B are one third and unity respectively. Neither assumption need be true since A will only be one third if the soil responds elastically and isotropically and B will only be unity if the soil is saturated. In desiccated ground the latter, if not both assumptions, is certainly in question.

While Driscoll is aware of the problems associated with swelling tests, he does not point out that exactly the same problem occurs when relating moisture content to soil suction via the liquid limit, as illustrated in his Fig. 4. His suggestion that desiccation is implied if soil samples plot below the pF = 3 line is not necessarily correct, for the same reason that a swelling test taken in isolation does not necessarily indicate desiccation, i.e. an overconsolidated clay may possess a moisture content much lower than a clay of similar index properties which has not been subjected to such a high overburden pressure in the past. An example is given in Fig. 3, where samples of Lower Lias clay from open ground and from about $0{\cdot}3H$ from a row of lime trees are plotted. Sampling was carried out in late August 1983. Many samples from open ground fall close to or below the pF = 3 line, suggesting significant desiccation. Samples from close to the trees, however, fall very much further below the line, perhaps in the vicinity of pF = 4 at worst, though there is still some overlap in the results.

These results and comments highlight that the only reliable way to establish whether desiccation is present is to compare soil properties for samples taken close to trees with those taken in open ground. This point is not made strongly enough by Driscoll. If no opportunity exists for establishing an open ground control, then it is not possible to draw absolute conclusions about the degree or depth of desiccation attributable to the trees, unless local experience is available. A desiccated clay may be expected to possess a lower moisture content, but a higher consistency index, dry density and undrained shear strength than the equivalent clay in open ground. Swelling pressure and percentage swelling may also be of value but the relative expense of these tests precludes their use for small projects. As an example, Fig. 4a shows plots of moisture content, consistency index and dry density against depth for the same site as Fig. 3 where desiccation is implied to about 3m.

P. G. Biddle

Biddle presents some very interesting results regarding moisture deficiencies at varying distances and depths from a number of species. We feel, however, that the Paper is optimistic in its conclusions since the number of trees under study is very small, considering the number of variables that can affect the degree and extent of desiccation. There seems little recognition that hydrogeological conditions probably have as much effect on the extent of desiccation as the tree species itself. Although it is obviously important to recognize that different species have differing water requirements, the idea that the identification of the species alone can also define the approximate size of the potential zone of desiccation should not be promoted. The extent of the desiccation zone will depend largely on the prevailing geology, topography, hydrogeological and meteorological conditions. The examples in Fig. 4, which show the influence of lime trees and small fruit trees extending to about 3 m on Lower Lias clay, demonstrate that trees that are generally regarded as low demand may have a serious effect in some cases.

Most of Biddle's results of moisture deficiencies relate to a comparison between autumn and spring. It is not surprising, therefore, that the most significant changes in moisture content occur in the upper 1·5 m, since this is the zone in which rehydration throughout the winter

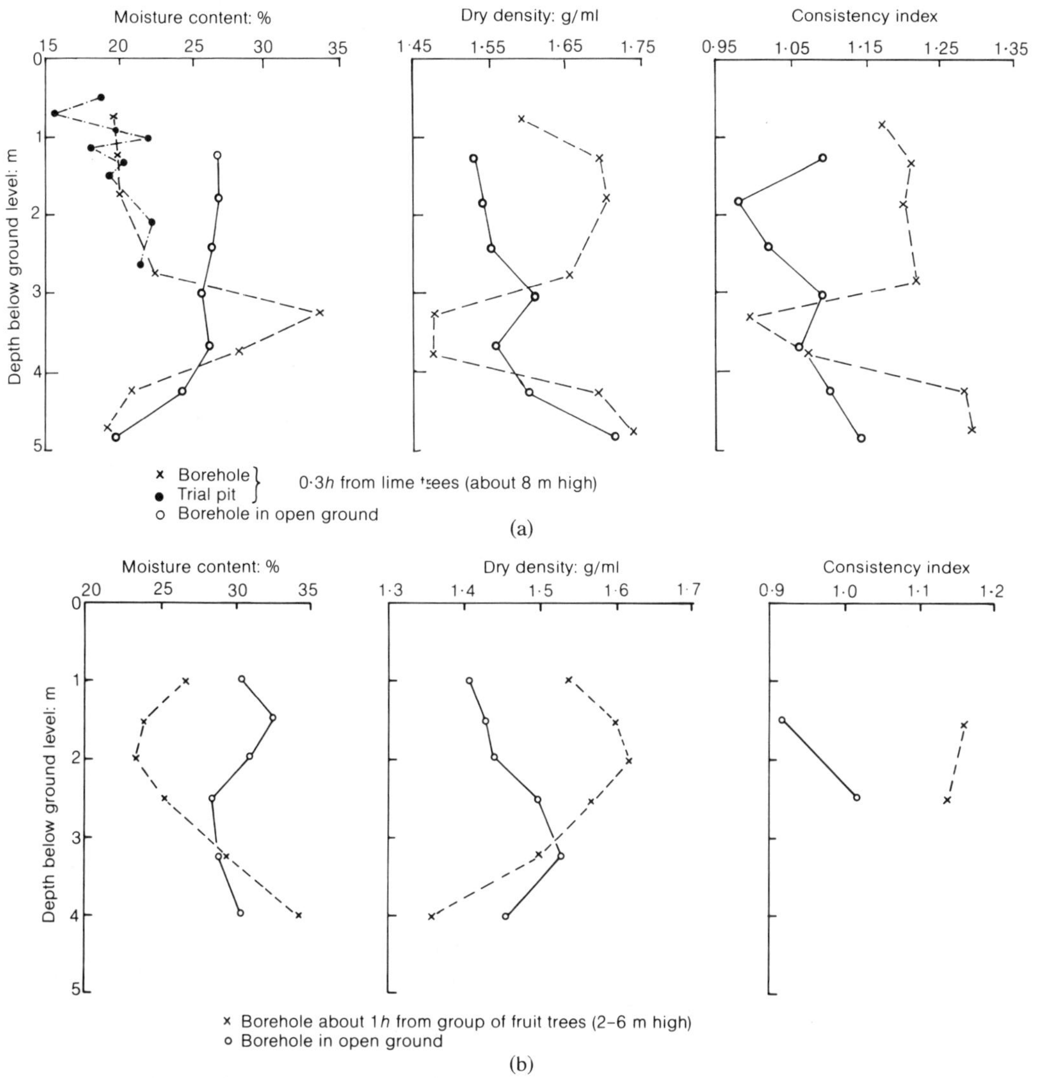

Fig. 4. Influence of trees on two Lower Lias clay sites: (a) site influenced by lime trees: (b) site influenced by small fruit trees

might be expected. The results shown presumably refer to established semi-mature or mature trees where the growth rate is fairly stable, and one may expect that the lower demand trees do not appear to cause additional desiccation through a series of near normal summers. We should perhaps remember that many buildings survived undamaged for years, adjacent to established trees, until the drought of 1976. This drought of course changed the hydrogeological conditions, as well as increasing the demand for water by the trees, thereby extending the degree and zone of desiccation. We consider therefore that Biddle is premature in suggesting that we only need to have regard for high demand trees, unless the Engineer is only required to design against average conditions and not the 30 year (or even more infrequent) drought. If we are to design against such drought conditions, where should the investigator look for his guidance upon depth and degree of desiccation? After all we cannot ask clients to defer the investigation until the end of the next drought summer.

D. W. COX (Polytechnic of Central London)
R. Wyles

Wyles suggests that any professional man ignoring the conclusions of BRE Digest 251

(Building Research Establishment, 1981) may be at risk. One of the central features of the digest is a guide to the structural importance of various amounts of damage, which is very useful and of considerable help in reducing unnecessary and expensive repairs. In particular, crack widths are given degrees of importance and hence necessity of repair. However, what is not fully expressed in the guide is the degree of risk or uncertainty attached to the structure, which is the underlying cause of loss of value, and the law relates to loss of value rather than structural engineering.

For example, a crack width of 1 mm per metre may be considered negligible in a mining subsidence area or in a house built with cement based bricks that might shrink by that amount. The same crack width may be of minor concern in a house on clay near trees. However, if a house ostensibly on dense gravel was cracked to this degree then the cause is probably serious and collapse may be imminent.

I therefore suggest that before recommending such strong adherence to Digest 251 it is revised to allow more fully for the circumstances such as soil type, location and fabric materials, which at present are only briefly mentioned. Otherwise, the recommendations can be possibly misleading when applied to circumstances other than clay brick houses on clay soils.

P. G. Biddle

I would like to comment on Biddles' conclusions regarding the development of a permanent moisture deficit. The experiments described register annual seasonal changes in moisture content and not the total deficit. The results show annual changes around an established tree and do not show the effect of planting or removing a tree. The conclusions, however, refer to permanent deficits, which were not measured.

There are references in the literature to seasonal ground movements around trees of 25–50% of total cumulative movements. For example, the BRE published measurements typically showing 25–50 mm seasonal movements and 50–100 mm total movement on removal or planting of a tree, although not necessarily on the same sites. On this approximate basis Biddles' estimates should be at least doubled to give total movements.

Permanent deficits can often be shown by moisture content measurements at various depths and distances from trees, on samples taken by pits or boreholes. The samples should be checked for changes in soil type by standard classification tests, but if the soil is laterally consistent, then any lateral changes in moisture content are due to the trees.

Figure 5 shows results of moisture content and plastic limit tests recorded this year, of a 20 m tall ash tree in Regents Park, where the desiccation extends 10 m horizontally and 4–5 m vertically. This is based on about 300 moisture content and 30 plasticity tests.

The distance at which damage occurs has already been well identified by Ward (1953) and Cutler & Richardson (1981) among others.

The depth at which the desiccation deficit ceases, i.e. the depth at which foundations must be placed to avoid movement, is the major uncertainty. To determine this an economic method used by the writer is to dig test pits at various distances from a tree on a radial line and then sample both ends of each pit to give a detailed moisture profile at several positions.

May I suggest that if this method of measurement was a standard procedure and Driscoll at the BRE was able to accumulate the results, a pattern would soon emerge as a result of co-operative effort.

I. Ravina

With regard to the influence of cracks on soil movement I would confirm Ravina's observations that cracks only go to shallow depth and terminate in slickensided fissures at 45° inclination. I have carried out examinations at several sites with similar conclusions. In some cases shrinkage cracks extend right to ground level at an inclination of about 45°.

The conclusion is that overconsolidated clay consists of individual blocks divided by slickensided fissures at about 45° inclination. Individual blocks can then expand or contract as necessary with the volume change taken up by sliding on the slickensided fissures. In this way it is possible to produce vertical surface motion without consequent deep vertical tension cracks which should theoretically occur with the concurrent horizontal shrinkage.

The expansion (or contraction) is shown on Figs 6–8 together with vectors showing the motion of each point. Two consequences of this are that lateral motion of clay soils occurs at the boundary of moisture change and that it is a function of the depth of and amount of moisture change.

Minor lateral extension of structures on clay soil is relatively common. More severe damage may be caused by repetitive motion which may progressively widen a crack during swelling but not close it so easily during shrinkage.

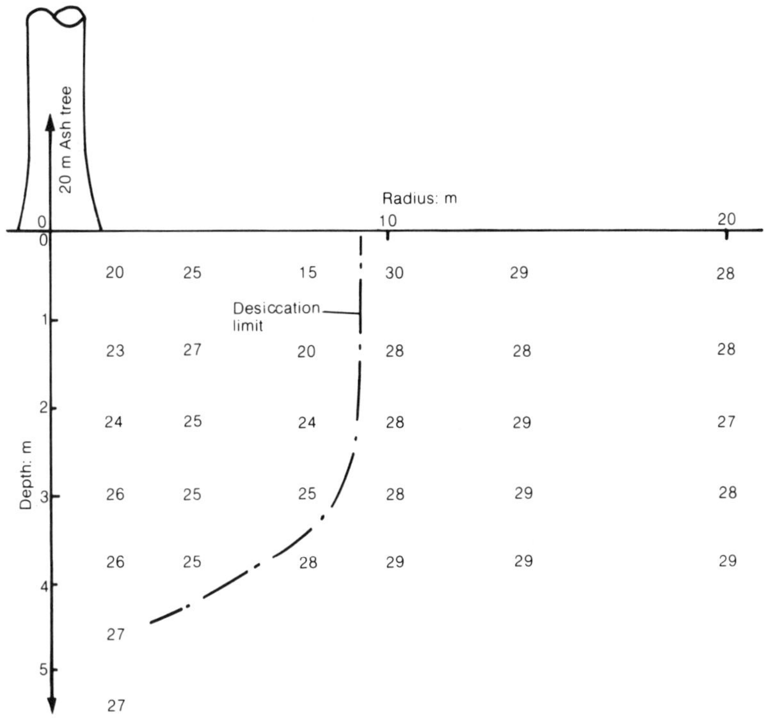

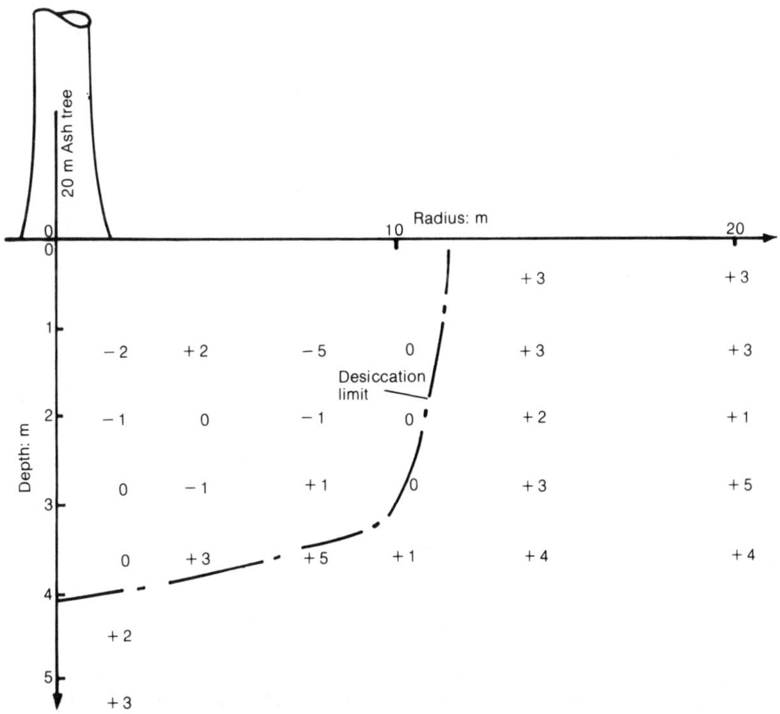

Fig. 5. Ash tree on London clay, Regents Park: (a) Moisture content distribution (average 300 + measurements); (b) moisture content–plastic limit distribution (300 + m.c., 30 PL measurements)

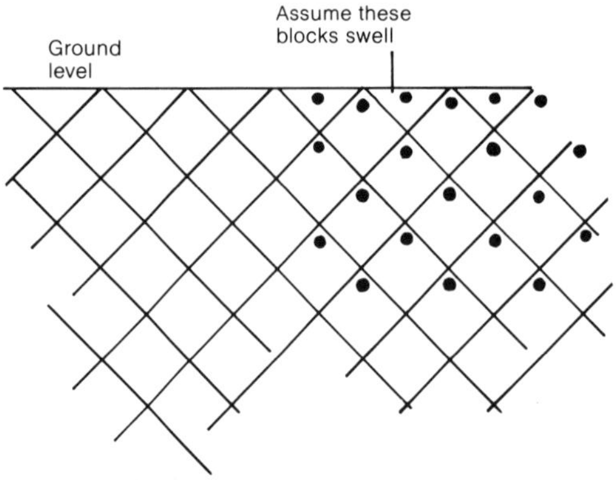

Fig. 6. Observed crack/slickenside pattern

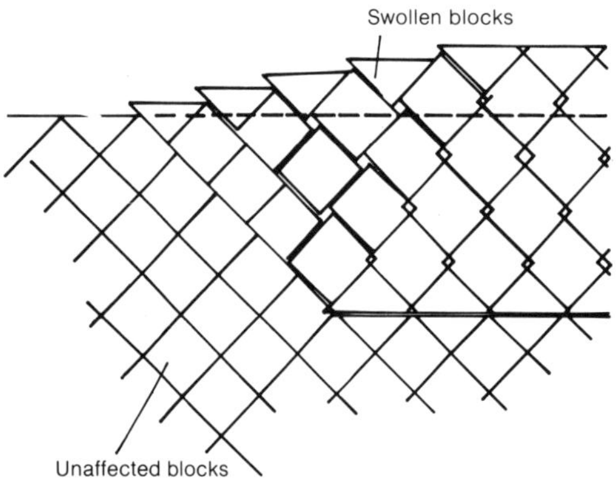

Fig. 7.

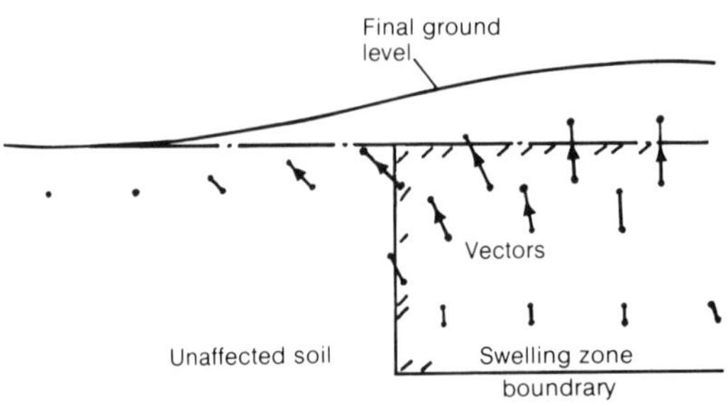

Fig. 8. Vertical and lateral motion

General discussion

The Papers refer to suction or negative water pressure measurements as the key to determining the effect of vegetation. There are other engineering techniques that can be applied as a consequence of knowing the negative pore pressure. One which is at present being examined is to use a flownet (or similar mathematical flow models) to establish the flow from the water table to a tree (or line of trees).

If it assumed that the tree removes water in the same way as a pumped well or sump then the extent of the drawdown can be measured and the tree water demand (well production) determined by standard calculation methods. If the typical demand of a particular species can be established then it will be feasible to predict the drawdown curve for that species on any soil, provided the horizontal and vertical permeabilities are known. From the drawdown, the moisture changes and hence depth of foundation can also be predicted.

Additionally the model can be used to predict the effect of a root barrier or an irrigation trench and hence determine suitable dimensions.

As with well calculations the effects of variable demand can be simulated in the mathematical models or the demand averaged for a flow net.

N. C. TRAIN (Baynham Meikle & Partners)

The site that poses the greatest dilemma for the designer is one that has a clay of high shrinkage potential, but which is devoid of trees or other vegetation to readily justify enhanced foundations.

The problems that trees could cause on such a soil are well documented and, moreover, in BS 5837 (clause 8.4.2.2; British Standards Institution, 1980) it states that 'where roots have been found or are anticipated in the future, it is advisable to take precautionary measures in the design of the building foundation'.

Although in the code it also counsels (clause 8.4.1) that, in buildings that have not been designed for the effects of shrinkage and heave, there is a possibility of future damage if trees are planted too close, it is unlikely that the householder will be aware of such a warning. If the foundations do not include for shrinkage, it is possible that the householder, in planting a windbreak of, say, poplars at the back of his patio, may cause a shrinkage problem during the life of the building. Since the damage would date from some time after the planting of the new trees, and not the date of construction, the designer could find himself accused of negligence long after construction.

Hence, with a highly shrinkable clay, the designer on the one hand may lose a case of negligence if he does not provide enhanced foundations, and on the other may lose a client if he does not provide economic foundations on a treeless site.

A possible solution to this dilemma would be to provide shallow foundations, disregarding the effects of shrinkage, and then ensure that both the builder and the householder were aware of the problem that future tree planting could cause. NHBC Practice Note 3 (National House-Building Council, 1969) gives clear guidance as to the positioning of trees and can be understood by the layman. If this note was appended to the deeds of the property and it was made clear that no responsibility would be taken by either the builder or the designer for damage caused by trees planted closer than recommended for a building with foundations at 1·0 m depth, would this be sufficient to absolve the designer of responsibility of negligence? And would that absolution be maintained if the house was resold?

B. A. LEACH (Allott & Lomax)

R. Wyles

Many houses were built in the 1950s on clay now recognized as prone to shrinkage. In many cases these houses were built on estates where trees had already been planted along road verges. At that time it was not general practice, as it is now, to take foundations deeper in such clays and many were constructed to the recommended 0·9 m. In the case of subsequent damage attributable to the trees it seems to me there are two arguments. Either the house developer is to blame, despite common practice at the time for not providing adequate foundations, or the owner of the trees is to blame because they clearly have caused the damage, even though they were there first. It has been suggested that the latter interpretation would apply. What would be the legal view in such a case?

R. Driscoll

Driscoll quotes his Table 2, which gives shrinkage potential in terms of plasticity index and clay fraction, from BRG Digest 240 (Building Research Establishment, 1980a). The problem with this table arises when in 'lay' hands. Simply because a clay falls into the table even on a low potential rating, separation distances between tree and property, of 1h or more begin to be bandied about. Clearly there is, or ought to be, a scaled distance for similar trees in clays of differing potential. Now that we have some quantitative evidence of soil moisture deficit for differing species of trees is it possible to grade the clays in Driscoll's Table 2 by allocating appropriate water shrinkage factors to each category? Perhaps the Author could suggest values for these.

With regard to heave, I have a case where quite noticeable cracking during the 1975–76 drought is reported to have completely closed during the following winter period. I would welcome the Author's views as to whether shrinkage is 100% recoverable or whether it is analogous to the consolidation–swelling situation where, commonly, swelling movement is less than consolidation movement for similar load increments. On a similar point, I have heard it argued that swelling due to removal of trees will not lift shallow foundations because of lack of confining pressure and yet there are a number of such cases where upward movement has taken place.

T. SHARROCK (Liverpool Polytechnic)

In referring to the detection of desiccation Driscoll implies that soil test methods are outside normal site investigation practice. I am surprised that more use has not been made of the shrinkage limit test (TRRL method). This is a versatile and rapid test that not only gives the shrinkage limit but also the shrinkage ratio which will indicate the amount of shrinkage likely to occur when water is removed from the soil by vegetation.

Of more importance, however, are tests to determine the suction characteristics of soils. These have been developed by soil scientists and such methods as the glycol osmotic method or filter paper method are simple, economic, rapid and readily available for the determination of the suction characteristics of soils. These methods have been used successfully at Liverpool Polytechnic in the investigation of properties of clays in north-west England.

This soil suction approach is an important technique in investigating the shrinkage and swelling of clay soils.

R. D. SEAGOOD

Loss adjusters are heavily involved with claims for subsidence and heave damage to domestic structures for the insurer or underwriter who has issued a policy to the householder, a party alleged to have created a nuisance (tree owner) or a practitioner who has been held negligent.

Adjusters strive to ensure that adequate investigation is carried out into causation but are hampered because, frequently, surveyors and structural engineers do not fully understand the clay subsoil that they are dealing with, rarely can distinguish between quaternary deposits and overconsolidated clay and do not possess the equipment to carry out the simple in situ testing. Moreover, the taking of levels as recommended by the Building Research Establishment is perhaps the most significant omission, quite apart from the fact that little or no attention is paid to the classification of physical damage. As a result, reports are issued to a worried owner, advising that extensive underpinning is necessary and that a guarantee should be obtained without even a trial pit sometimes having been excavated or the geological map consulted. In addition, there is a reluctance among professional men to amend their preliminary report even when further information becomes available at a later date.

Against such a backcloth it is not perhaps surprising that it is discovered, occasionally, that the crack damage is due to roof spread, thermal cracking, failed lintels, incorrect use of calcium silicate bricks, etc. or that there virtually is no differential movement at all. It is necessary to maintain reasonable cost control, consistent with a sympathetic understanding of the insured's problems, and a necessity to be realistic and make decisions pragmatically when appropriate. However, it is not unknown to receive schemes designed by geotechnical specialists after detailed ground investigation, that can only be described as an 'overkill'.

On a fairly level site, i.e. not subject to the effects of creep, damage by clay shrinkage within category 1–3 often can be dealt with successfully by removal of the tree or trees identified to be involved. Nature can be harnessed to effect recovery which can close summer cracks from 0·25 in. to hairline the following spring, thus relieving the stress in the building and reducing the distortion. Clearly, claims about continuing heave after the removal of trees are exaggerated, but that does not mean that one should not be cautious about removing a tree that is older than the structure, particularly when the tree is very close and the building is founded on a clay with a high liquid limit. Nevertheless, where trees have been removed and the clay subsoil is allowed to recover the degree of success is remarkable, whereas deep partial underpinning may create a potentially worse situation for some luckless future owner or insurer. Massive repairs, namely underpinning and the installation of fully suspended floors, have been avoided in heave cases where again the damage has not been too severe, merely by monitoring the building and effecting repairs to the superstructure when the building has settled down.

While many insurers have recently increased rates on domestic properties, this has been coupled with the introduction of cover for damage by heave of the site and a reduced policy excess. As Driscoll points out, the current situation reflects an over-emphasis on underpinning and I would advocate that case studies are brought to the attention of a much wider audience in the professions to promote a better understanding of what can be achieved with common sense, care and expertise, together with a willingness of all concerned to learn new techniques, not to be too inflexible and not to regard underpinning as the only remedy available, which it certainly is not.

E. J. WILSON and M. C. BURBIDGE (consulting engineering geologists)

R. Driscoll

Driscoll identifies soil suction as the fundamental parameter of desiccation, but recognizes the need for simpler routine identification tests. As rough approximations he defines significant desiccation (pF = 3) in terms of moisture content, on the one hand as 0·4 × liquid limit, and more tenuously on the other as 2% above plastic limit. These two concepts are mutually exclusive except for a unique plastic range. This problem can be overcome by the use of the more rational dimensionless parameter, consistency index, defined as

$$I_c = \frac{LL - w}{LL - PL}$$

Consistency index equals unity at the plastic limit, with increasing values as the soil becomes more desiccated or overconsolidated. Data from Driscoll's Table 7 gives average values of I_c = 0·8 for pF = 2 and I_c = 0·9 for pF = 3.

There is a serious deficiency in attempting to define desiccation by any parameters in isolation. Most desiccation or heave damage is associated with overconsolidated soils, which at greater or lesser depth assume moisture contents and consistency indices akin to those produced by desiccation. It is therefore essential, in order to separate the effects of the two, to include in every investigation a control set of measurements from an additional borehole or pit in otherwise similar ground conditions remote from the suspect vegetation, and to base assessment of desiccation on a comparison between the two.

P. G. Biddle

I congratulate Biddle on the systematic approach he is making towards determining soil moisture deficit patterns, but find it sad that we have had to wait 35 years from the initial work by Ward (1953) and by Croney & Lewis, to begin the long and painstaking process they urged then.

In view of Wyles' closing sentence, however, I want to emphasize Biddle's reservations on publishing his work at this stage, since at first sight his Paper seems to present exactly what we have all been waiting for. It is not complete. He himself states that observations must be carried through a drought period before they become meaningful, and that means before they can be used in foundation design. With an experienced periodicity somewhere in the order of 30 years, severe drought is almost certain to occur at least once during the lifetime of every house currently being built. Biddle has also warned that the results so far published relate only to annual soil moisture deficit and not to cumulative deficit. Again, this seriously limits their practical usefulness at this stage.

It is unfortunate that Biddle's work has not included a study of oak or willow, since oak is now emerging as statistically even more damaging than poplar.

The early work by Ward (1953) was based on ground settlement, from which soil moisture deficit and rooting patterns were inferred. Recent work by Cutler & Richardson (1981) presented a statistical survey of structural damage in relation to distance from trees, but with no detailed reference to soil or moisture deficiency. Biddle's work is based on soil moisture deficit and neither rooting pattern nor ground movement are measured.

It would be even more useful to see the results of a fully integrated programme which for each tree studied includes the measurement of both vertical and horizontal ground movement at surface and in depth, in combination with measurements of soil moisture deficit and the extent of root growth. It would even be useful to see confirmation of the tacit assumption that root spread and soil moisture deficit are equal in all directions from an isolated tree.

This data must be assembled for a sufficient number of trees in normal open situations to establish a clear norm, and must also be extended to cover some of those abnormal situations (e.g. competition with other trees, trees adjacent to covered areas) that will otherwise continue to defy the guidelines we are trying to draw up.

Clearly the above catalogue is beyond the present funding of Biddle's project, but the widespread interest he has helped to generate in this topic ought to encourage other research bodies, such as university and polytechnic departments, to pursue their own research, and perhaps each to adopt a tree or set of trees in their own area, for study. For the resulting data to be of the greatest use it would have to be collected to a standard specification. This specification should include detailed species, with an arboriculturist's report on condition and state of growth, geological classification of subsoil, engineering profile of subsoil (index properties against depth, groundwater levels), and should include the ground movement as discussed above, soil suction, moisture content and soil moisture deficit as reported by Biddle. If a neutron probe is used, then results should be correlated for each site with moisture content determined traditionally, so that the measured moisture content may be related readily to index properties. These measurements should be taken on more than one and preferably four radii from each tree. This work should be coordinated by an overseeing body, such as the Building Research Establishment, and should be started now, to avoid the risk of missing yet another drought.

Only in this way are we likely to achieve a sufficient depth and breadth of data to produce guidelines that can be used either for the design of foundations, or in the investigation of structural damage.

C. G. HOSKINS (Rofe, Kennard & Lapworth)

The effects of vegetation on clay soils were discussed with regard to structures but not to slopes, where problems may also arise following the removal of vegetation and the subsequent swelling and increase in pore pressure.

Vegetation will usually grow faster than long term pore pressures can develop in a newly formed clay slope, and thus the vegetation may maintain the stability of an otherwise potentially unstable slope. Problems generally arise when the vegetation, particularly trees and the large bushes, are removed at a later date and the soil moisture deficit is lost as the clay swells with the development of the long term pore pressure distribution.

This problem has been apparent on the downstream face of some old earth dams where removal of existing trees has been recommended in the past following a statutory inspection under the Reservoirs (Safety Provisions) Act. Removal appears to have been recommended for fear of damage to the core by root growth in search of water, although this generally has only occurred with long term lowering of a reservoir and additional growth of existing vegetation towards the lowered water level. Following removal of the vegetation, instability of the downstream face has subsequently become apparent as the pore pressures have increased beyond critical values.

An example of this behaviour occurred at a reservoir in Hertfordshire where trees, up to about 10 m high, were growing on a nominally 1 in 4 downstream slope of an eighteenth century dam built of weathered London clay. The reservoir had been inspected and a recommendation made that the trees should be removed, despite the history of slipping of the embankment from its earliest days. The trees were removed in 1973 and other minor works carried out. Some few months later two separate slips developed in the cleared upper levels of the slope, whilst the steeper lower levels which were left wooded were unaffected. The slipped areas were later dug out and replaced by gravel fill but continuing movement of other adjacent lengths of the shoulder led to the installation of 5 m deep drainage trenches down the slope over the affected length of the upper levels of the shoulders in mid-1983. To date no further movements have been noted.

Another aspect of vegetation on slopes concerns the increasing, and an environmentally welcome, planting of trees alongside motorways and major roads. Whilst the additional tree planting is to be applauded, the long term implications for the slopes when the trees die or are removed and for future maintenance should always be borne in mind at the time of planning and construction. Perhaps the selective pruning techniques discussed may be applicable in this instance.

M. G. ANDERSON (University of Bristol) **and P. N. KNEALE** (Trinity College, Dublin)

Critical moisture content for the onset of desiccation

The identification of the moisture content (or soil suction) at which desiccation in clays may begin is, as Driscoll observes, generally outside the scope of standard site investigation work. Accordingly, Driscoll has suggested that a plot of moisture content at 10 kPa soil suction (an assumed air entry value) against liquid limit, may prove a useful guide, with desiccation expected at moisture contents below this critical value. Work in this general field is inhibited, as Driscoll observes, by the lack of data available on the suction characteristics of British clays. This discussion extends the work reported by Driscoll in relaton to desiccation crack initiation.

Study site with automatic soil suction recording

An automatic Scanivalve tensiometer system facilitating the continuous recording of soil suction and positive pore pressures at 24 individual points, was installed for two years (see Anderson & Kneale, 1980a for instrumentation details). The site was a 26° ground embankment on the M4 motorway, being constructed of Oxford clay (liquid limit 73%). The instrumented area was approximately 30 m long and extended the full height (10 m) of the embankment. A detailed crack survey was made in a 12 m by 9 m

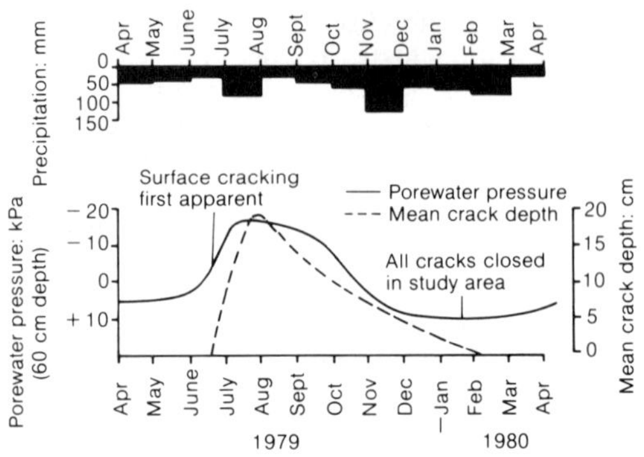

Fig. 9. Soil suction change in embankment of Oxford clay at 60 cm depth (M4 Wootton Bassett; GR SU 053836)

area and soil water suctions were recorded at depths down to 1 m, with the automatic equipment described. Fig. 9 shows the smoothed suction response throughout the year at 60 cm depth. Visible indications of crack formations were first observed in June, while subsequent closure was apparent in all locations by the following February. While crack initiation was observed to correspond to an average soil suction of 6 kPa (approximately the air entry value for the clay), subsequent closure was a function of the rainfall pattern—Fig. 5. However, of particular interest was the permeability of the intact (uncracked) zones, and those locations that had exhibited cracking during the summer. Following the surface closure of the cracks in February 1980, field permeability tests revealed surface permeability values of $5 \times 10^{-5}\ ms^{-1}$ for former cracked locations and $4 \times 10^{-7}\ ms^{-1}$ for intact sites. In addition, a number of 5 cm dia. core samples were taken and subjected to X-radiography (Anderson, 1979). These samples were taken from sites that had cracked, but had subsequently closed. Fig. 10 shows selected slices of a core subjected to X-radiography at up to 18 cm depth—the position of the crack shows up as a band of dark (less dense) material running across the lower half of each sample. At the surface, the crack was not visible, nor did such a banding show on the X-ray. The clear implication here is that the cracks, although closed at the surface, remain in terms of zones of significantly higher permeability below surface. This fact accounted for the relatively rapid response of the embankment porewater pressures to rainfall (see Anderson & Kneale, 1980b). Moreover, the crack locations at this site were apparently permanent features, not only in the temporal sense of the permeability change, but also in the spatial context—the 1980/81 season showed the

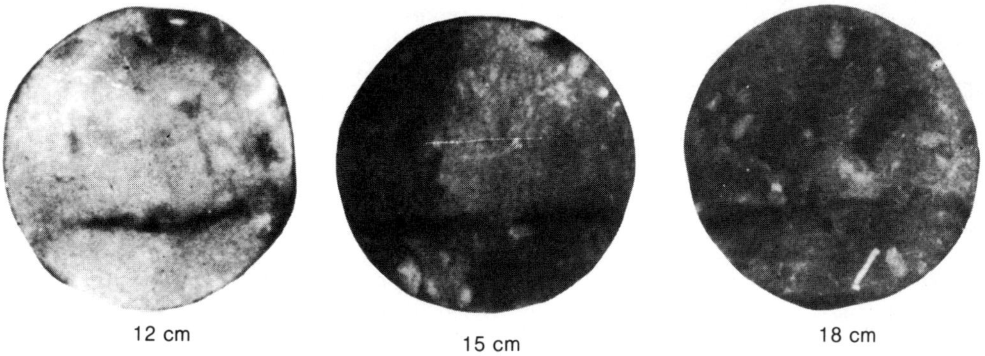

Fig. 10. X-radiograph of clay slices at 12–18 cm depths, taken from a site at which the crack was closed at the surface

same locations to exhibit cracks. Thus, while some support may be found here for Driscoll's suggestion of crack formation occurring at around the air entry value of clays in the context of visible surface effects, I suggest that the effects of such cracks may be present all year in rendering the embankment to be highly responsive to rainfall, even though in late winter and spring, the cracks are closed near surface.

J. M. McENTEE (Wimpey Laboratories Ltd)

I would like to comment upon the special foundations and their performance that have been adopted by my parent company, subsequent to the dry summers of 1975 and 1976, for housing units on shrinkable clays that are sited within the zones of influence of the root systems of trees.

Three types of foundation have been used—deep strip footings, pile foundations and reinforced concrete slabs with downstand edge beams constructed upon a granular blanket.

The deep strips are lined with expanded polystyrene to minimize the effects of horizontal and vertical swelling of the clay and a suspended ground floor construction with a 150 mm deep void beneath is adopted.

Pile foundations were generally formed by sinking an unlined borehole by means of a simple auger rig. The top 1·5 m of the pile was sleeved in polythene or the length of the pile was increased to resist uplift, and the ground beams were cast upon a 150 mm thick layer of expanded polystyrene.

There have been three instances of damage to housing units founded on deep strip footings on London clay and Keuper marl at depths of 1·5 m and 2·0 m and through fill to a depth of 4·25 m into underlying chalk. This type of foundation appears to be intolerant to movement and has now been modified to comprise mass concrete surmounted by a reinforced ring beam.

Uplift of short-bored pile foundations has occurred at two sites underlain by London clay in Essex, and current design of such foundations makes an allowance for swelling forces acting on the top of the pile shaft and also on the underside of the ground beams.

There have been no reported instances of damage to houses supported on slab foundations.

I conclude by giving case histories of the effects of uplift due to the swelling of London clay. The first site is in Hampshire where a

Levelling station	Difference in level of D.P.C. relative to station B: mm		Recovery of A: mm
	February 1972	September 1978	
A	3·5	0·2	87
B	–	–	78
C	73	1·5	7
D	65	3	16
E	22	11	66

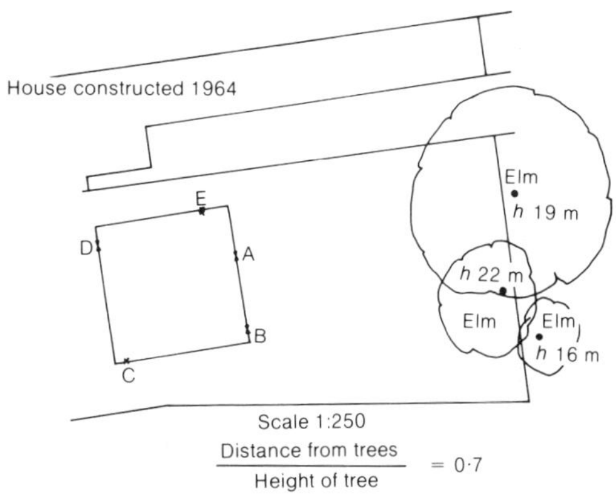

Fig. 11.

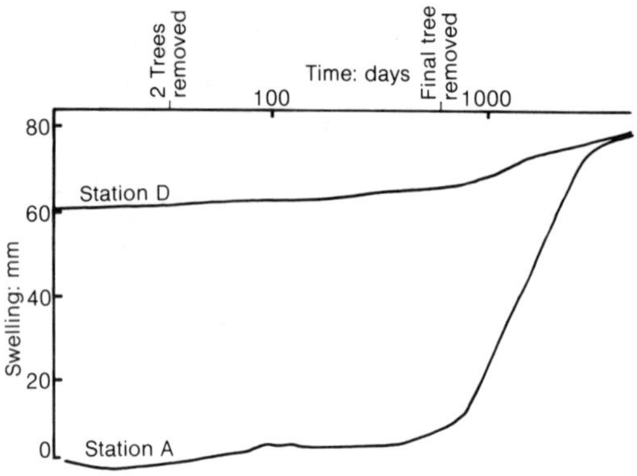

Fig. 12. Swelling of London clay due to removal of trees

detached house constructed in 1964 on 0·9 m deep strip footings was located near a clump of four elm trees the nearest being 22 m high at a distance of 13 m from the front of the house.

Damage was first reported in September 1969 which became worse in September of the following year. Two trees were removed in March 1971 and by November 1972 swelling of the underlying clay became evident. The remaining trees were removed in March 1976 when bad cracking including damage to the ceilings was reported. The results of levelling surveys indicated that the heave of the clay was almost complete in November 1977 with a maximum heave of 87 mm. Desiccation was indicated to a depth of 2·5 m. Figs 11 and 12 show the results of the survey. The damage to the foundations was confined to a single crack at the front of the house.

At the second site in Essex a terrace of four two-storey houses was sited 12 m from two 11·5 m high oak trees located on the bank of a small stream, which as a planning requirement was to be placed in a culvert. The layout of the site is shown on Fig. 13.

Construction commenced in late 1976, the units being founded on 1·5 m deep strip footings, with a precast ground floor construction and a 150 mm deep void beneath. The stream was culverted prior to construction, but only within the confines of the site to the manhole shown in Fig. 13.

The houses were occupied in the spring of 1977 and cracking to the front walls of nos. 2 and 3 was reported in the autumn of the same year. In February 1978 a severe storm resulted in the still open section of the stream overflowing and flooding the front gardens of the terrace.

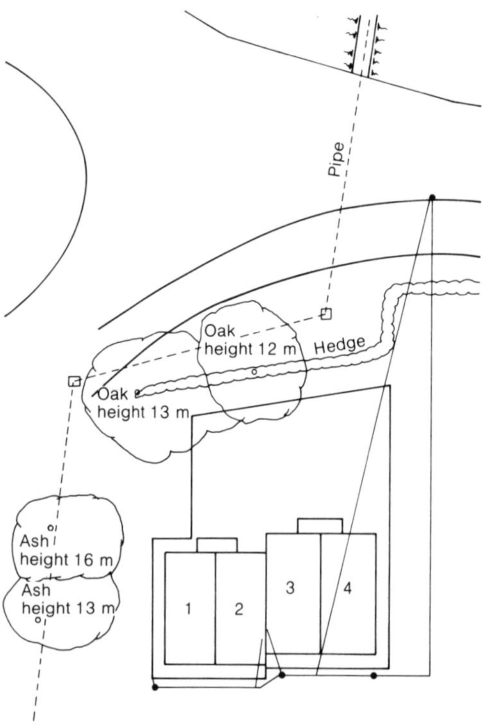

Fig. 13.

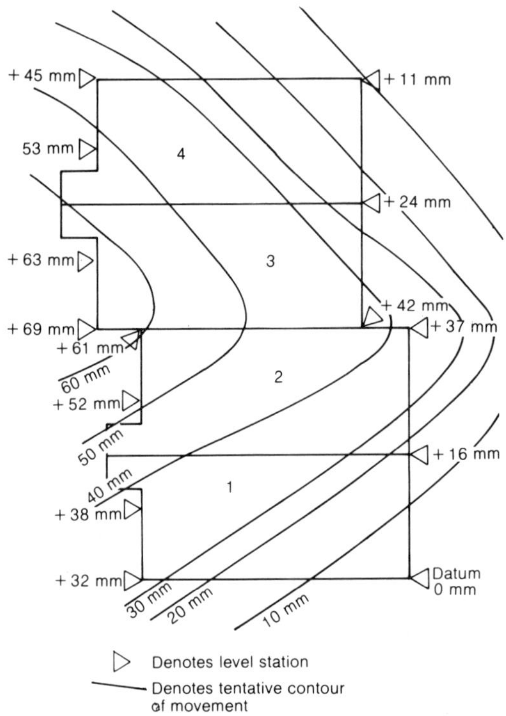

Fig. 14.

Inspection revealed that the foundation of the party wall between units 2 and 3 had moved 10 mm away from the front wall foundation of no. 3. This movement was probably due to the action of the root systems of the two trees being deprived of their source of water by the culverting of the stream. The subsequent ingress of flood water through the gap in the foundations resulted in heave of the underlying clay. The results of level surveys indicated that this movement had virtually ceased by the winter of 1982, with a maximum uplift of some 70 mm occurring below the party wall of units 2 and 3 and a horizontal displacement of 30 mm at first floor level. The contours of the uplift movement are shown in Fig. 14.

M. J. REEVE (Soil Survey of England and Wales)

A study of the shrinkage characteristics of various clay soils was carried out in the late 1970s by the Soil Survey of England and Wales. The results of our work on total shrinkage potential (Reeve, Hall & Bullock, 1980) largely agree with the data presented by Driscoll in Table 3. In addition, we recorded some extremely high potentials in Great Oolite (Blisworth) clay which had plasticity indices up to 78% and a potential to shrink to almost half of the initial, field moist, volume. High potentials were also measured in certain alluvial clays. Fortunately for the construction industry, Great Oolite clay is inextensive and sparsely built on and clayey alluvial soils, though fairly extensive (2–3% of England and Wales), are seldom built on because of flood risk.

The ability to predict accurately shrinkage potential from other parameters is clearly important. Table 2 of the Paper goes some way towards providing guidelines though I would question a reliance on clay content. Our findings (Table 1) were that clay content is not particularly well correlated with shrinkage potential. Clay percentages greater than 90% are almost unheard of in Britain though very high shrinkage potentials are encountered! The plasticity index or liquid limit should be preferred for predicting total shrinkage as they are both well correlated with potential shrinkage. An alternative and rather quicker predictive method, though only accurate when the soil is fully wetted (e.g. January to April in most years in Britain), is to core the moist soil and determine the dry bulk density. This parameter is also strongly correlated with potential shrinkage, low density clays having the greatest potential.

Table 1. Correlation coefficients (*r*) between potential shrinkage and various parameters for 42 clay subsoils

Parameter	Total potential shrinkage	Potential shrinkage 5–1500 kPa
Dry bulk density at 5 kPa	−0·86	−0·61
Clay (<2 μm)	0·64	0·29
Fine clay (<0·2 μm)	0·56	0·26
Plastic limit	0·61	0·39
Liquid limit	0·83	0·49
Plasticity index	0·83	0·47
Organic matter	0·56	0·23
Cation exchange capacity	0·63	0·23

Although such correlations encourage a degree of optimism for predicting shrinkage, in real situations a knowledge of total shrinkage potential is not sufficient. Subsoil clays in humid climates never dry sufficiently to attain their total potential for shrinkage. Thus it is important to relate shrinkage to annual moisture regimes and to the influence of vegetation and suction characteristics on these regimes. Data on suction characteristics are commonly collected in agricultural soil investigations, although it extends only to depths of 1 to 1·2 m below the ground surface. The Soil Survey of England and Wales alone holds a database which includes suction/moisture content relationships on 600 subsoil clay layers (clay>35%). For some of these, Atterberg limits and potential shrinkage are also available.

In assessing actual shrinkage potential, the use of the relevant suction limits is vital. Most clayey soils in Britain suffer seasonal waterlogging, most commonly due to perched water above an impermeable subsoil. In early spring in most years they are either saturated or at a low suction (1–5 kPa, pF 1–1·7) depending on site drainage. During the growing season, little water is lost by direct evaporation; most is through transpiration by plants and trees. As stated elsewhere in this symposium (Biddle; Williams & Pidgeon), in the absence of a barrier to rooting, different species exploit water to different depths in a soil, but generally only up to a suction of 1500 kPa, pF 4·2. *Thus it is the shrinkage over the 5–1500 kPa range that is relevant.* From our research, an alluvial clay with a high total potential shrinkage and a Keuper marl clay with a low total potential shrinkage have very similar shrinkage potential (5% and 4% linear, respectively) between 5 kPa and 1500 kPa suction. Unfortunately, shrinkage over this suction range is not well correlated with other physical parameters (Table 1) and can thus only be determined by detailed site investigation.

R. J. CHANDLER (Imperial College of Science and Technology)

I would like to congratulate Biddle on providing us with valuable data, unique in the UK, on the extent of the water deficit generated by tree root growth in clay soils. The value of this data would, however, be further increased if it were possible to add some detailed information on the ground conditions at the various sites. Whilst I appreciate that some geological and geotechnical data are given in the Paper, it would be valuable to have detailed soil profiles, including particularly soil descriptions, at least to the depth of influence of the tree roots. Soil profiles are rarely uniform with depth, and at the relatively shallow depths to which house foundations are taken, processes such as frost action, seasonal variations of soil moisture, former solifluction, hill creep, and so forth, produce near surface complications to the soil profile that no doubt explain in part some of the irregularities in soil moisture deficit reported by Biddle. Even a brief statement of the depth of the near surface disturbances of this type will aid the reader's interpretation of the data. Where glacial tills are concerned, with their well known, sometimes extreme, lithological variability, a detailed soil profile is essential for a full understanding of the soil moisture patterns that have been observed.

Tree root action in clay soils affects house foundations in two ways: shrinkage owing to root growth beneath foundations, and swelling where houses are built over ground previously desiccated by the root systems of trees now removed. In the latter case, moisture content observations by themselves are not sufficient to indicate whether or not the ground is likely to swell. The most satisfactory method of achieving the latter aim is likely to prove to be the direct measurement of the soil suctions that have been generated by the desiccation process, as described, for example, in the paper by Richards *et al.* Though there are not inconsiderable difficulties involved in so doing, I would like to urge Biddle to attempt such measurements if he becomes involved with further work of this type.

J. N. HUTCHINSON (Imperial College of Science and Technology)

The practice of root pruning received no attention in the symposium and was dismissed as of no value when asked about in the discussion. Would Biddle say this view is based only on hearsay or whether it has been substantiated by properly controlled experiments.

A. COATSWORTH (Soil Mechanics Ltd) **and J. EVANS**

Williams & Pidgeon have presented a comprehensive review of their subject making use of their wide experience and the large number of records held by the National Building Research Institute. The first writer's experience of heaving clays in South Africa is limited to just one site, but it was a large one and examined in detail, possibly more so than other sites in respect of this problem. The writers wish to highlight two areas of divergence with the Paper.

Firstly, on the depth of occurrence of roots, the Authors state that a review of common profiles in South Africa indicates that easily visible roots seldom extend below half the depth to the water table, and give examples of rooting depths, none exceeding 5 m. Gum tree rooting to 27·5 m is mentioned as an isolated record. The key phrase is 'easily visible roots'. There are few indigenous trees on the highveld, and eucalypt species, which are widespread throughout the country, were introduced around the turn of the century from Australia. Therefore, a large proportion of the trees have been planted. Excepting the tap root (which may be absent from a planted tree), the large roots of a tree are mainly there to keep a tree standing—they primarily fulfil a structural role, rather than collecting nutrients or water. Uprooted eucalypts frequently have large roots spreading out at shallow depths. However, in considering the effect of trees in desiccating the ground the interest lies in the thousands of very fine roots, not the 'easily visible roots'. The distinction is important when one considers that the vast majority of site investigations in South Africa is carried out by examination of the soil in situ by means of a large diameter (e.g. 750 mm) bore. This is an excellent technique, but care is needed to observe the all important fine roots under the poor illumination pertaining at depth. A rye plant is said to have roots with an aggregate length of up to 6 km! On the site examined by the first writer, fine eucalypt roots were observed to a maximum depth of 22 m, and in several cases these roots were below the current water table which is rising following removal of the trees. The writers suggest that careful study of rooting on many highveld sites would reveal that roots may go to much greater depth than indicated by the Authors, sometimes indeed to reach the water table, the level of which may be depressed by those very roots.

Turning to the question of heave prediction, the Authors advocate a relationship due to Brackley relating heave to soil suction, final vertical stress and plasticity index. Measurements were made of total suction on 19 specimens from the site investigated by the writers. On eleven of these a swelling test was also carried out. The sampling process might have allowed some wetting up of the samples, but the oedometer specimens would have been affected more than the small suction test specimens. Although Brackley's method predicted no swell for 10 out of the 11 specimens, the average swell from the swelling tests was 0·8%, not that small considering the range of depths (10–20 m) considered. An alternative swell prediction method, based on suction, due to Johnson & Snethen (1978) showed good agreement with the swelling tests. Furthermore, a nearby levelling point has risen 37 mm since the samples were taken, which was itself four years after the trees were felled. The failure of Brackley's method, when applied to a site on the widespread residual *Ecca series*, is important in discussion of South African conditions. The use of suction methods is likely to be semi-empirical, unless one determines the suction–moisture curve on a wetting cycle, and the first writer could find no laboratory in South Africa or the UK able to do this. It would seem that although research workers are keen to advocate the measurement of suction, this being the driving force in swelling, the prediction of swell is best based upon swell under load tests, which unlike suction measurements can be carried out in any commercial soils laboratory and do not rely on remoulded soil properties such as plasticity index.

J. M. RODRIGUEZ ORTIZ (Polytechnical University of Madrid) **and A. A. SERRANO** (Laboratorio de Carreteras y Géotecnia Madrid)

Large areas of Spain are covered by clays of montmorillonitic type, associated with an arid or semiarid climate, which leads to intensive swelling and shrinking phenomena. Fig. 15 shows the most endangered areas, which are generally covered by Miocene clays.

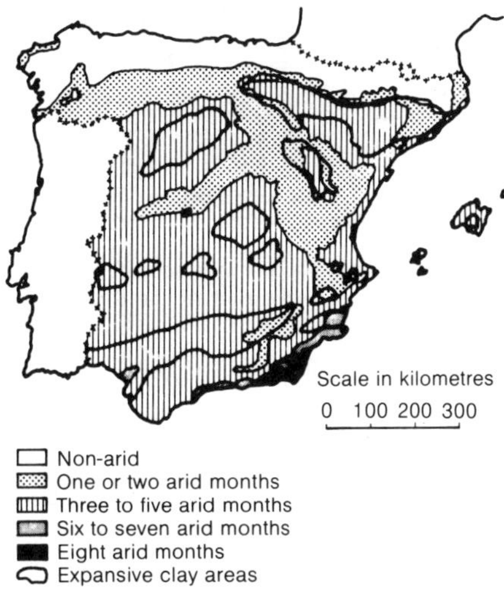

Fig. 15. Climatic conditions of Spain and distribution of swelling clay deposits

In rural areas local knowledge and intuition prevents damage from plant and tree growth near buildings, and problem trees are promptly removed at early symptoms of desiccation or cracking. In modern urban areas, however, gardening and tree planting has very often proceeded without special attention to these problems, with the choice of fast growing species of good visual impact, mainly of deciduous type, such as elms, poplars and acacias. No preventive rules have been followed with regard to, for example, distance from buildings and irrigation sequences.

A common feature of residential housing has been the location of shrubs along the external walls or in an open patio at the rear of the houses, with strong competition among neighbours about coverage by flowers and vegetation.

In the mid-1960s damage to a number of buildings was observed in a residential area in the south of Madrid. The ground consisted of montmorillonitic clay with the following properties: $w_L = 40–120\%$, $PI = 20–70\%$, $w_S = 8–30\%$, and activity $A = 0{\cdot}5–1{\cdot}5$.

The values of total soil suction ranged from pF 3 to pF 5, with remarkably high values above pF 4 very close to the artesian water table 9 m deep (Natural void ratio $= e_0 = 0{\cdot}6–1{\cdot}0$, c_c/c_s

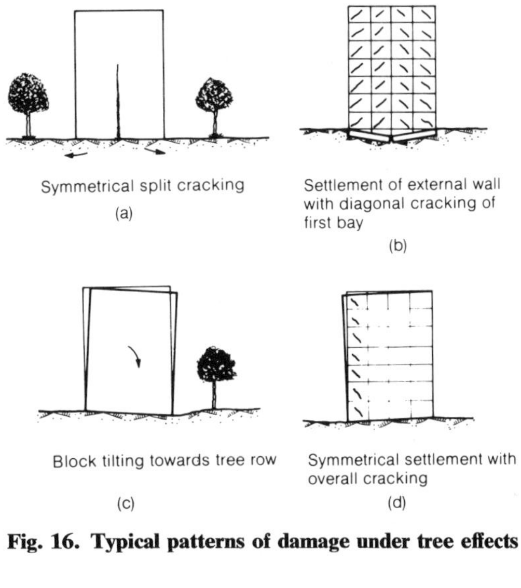

Fig. 16. Typical patterns of damage under tree effects

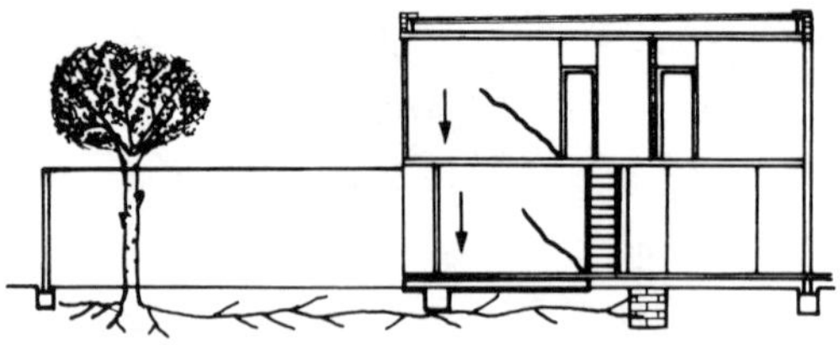

Fig. 17. Root penetration towards sewage sump causing wall cracking and settlement of the floor slab

ratio = 2–3, free swell = 1–7%; swelling pressure normally below 0·2 MPa).

The foundations of most of the damaged buildings consisted of concrete spread footings at very shallow depths, mainly 0·50 m even for four-storey buildings, with variations between 0·30 and 1·00 m. The spaces between buildings were left unpaved and poplars and elm trees were planted 1–4 m from the building and these had grown to over 12 m in only 5 years. Extensive felling carried out in the sixth year also marked an increase in widespread cracking.

The patterns of damage were of the following types (Fig. 16)

(*a*) Tilting of high buildings towards a close tree row, generally in front of the most sunlit wall
(*b*) Split cracking of front walls by shrinking of both sides of the building by tree desiccation
(*c*) Arch cracking above large roots penetrating beneath the building towards the sewage utilities. Very often the roots have forced penetration through the poor concrete of the footings and badly cracked the brick sumps or the concrete pipes, even appearing at the upper drain holes (Fig. 17).

Very little damage by trees occurred with footings exceeding 1 m deep.

An attempt was made to evaluate the risk of problems from trees through the following index

$$\mathrm{IP} = \frac{n_i A_i}{L} \delta \qquad (1)$$

where n_i is the number of trees of A_i grade summed for the length L of the most endangered external wall, A is an index depending on the species (see Table 2), and δ is a distance factor ($\delta = 1$ for $D < 5$ m, $\delta = 0{\cdot}5$ for $5 < D < 10$ m, $\delta = 0$ for $D > 10$ m, where D is the average distance between the tree row and the wall).

Table 2. Values of index A

Type	A
Elms	10
Poplars	9
Acacias	8
Willows, horse chestnuts, sycamores	6
Conifers	3

Table 3. Criterion of evaluation of risk

IP	Risk of damage
0–0·25	Very low
0·25–0·50	Low
0·50–1·00	Average
1·00–2·00	High
>2	Very high

Table 3 shows the criterion of evaluation of risk. This evaluation was applied to a group of apartment-type houses that had several degrees of damage, with fairly good correlation irrespective of the soil type (Fig. 18). Further refinement of equation (1) should include the soil swelling or shrinking potential, the foundation depth, the relationship between δ and the root length and the building capacity to bridge localized subsidence, i.e. the number of storeys or the type of structure.

Measures to remedy the damage to the apartments included

(*a*) replacing deciduous trees by perennial ones, usually conifers
(*b*) placing of insulating and root-resistant membranes, separating private gardens from nearby houses (Fig. 19)
(*c*) removing of shrubs and paving around the buildings.

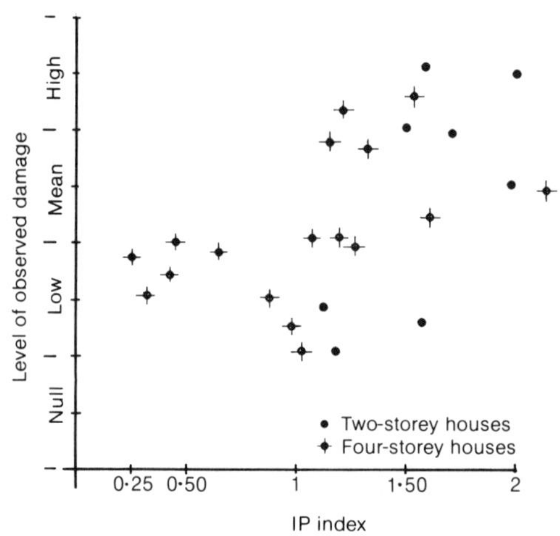

Fig. 18. Qualitative relationship between the observed damage and the risk index for the effect of trees

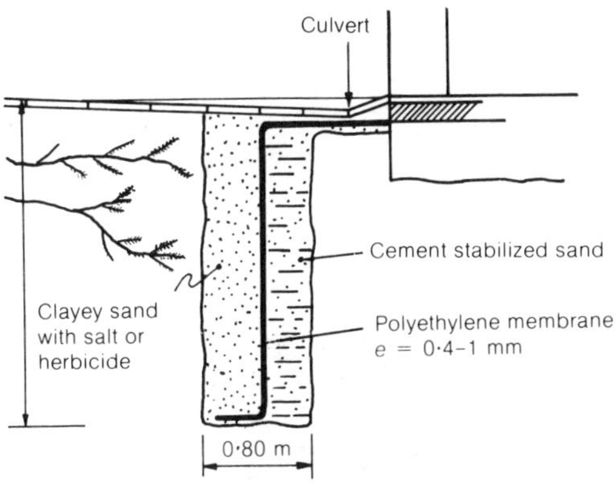

Fig. 19. Detail of ditch barrier against root or moisture penetration from backyard

The effects of desiccation have increased dramatically in years of exceptionally dry summers. Only in the Madrid area, have variations of Thornthwaite Index from −30 to −80 been recorded in successive years, with a mean annual rainfall ranging from 650 to 220 mm.

In a village near Madrid several dwelling blocks built 12 years ago have abruptly shown generalized damage this year (Figs 20–22). The two-storey houses, with brick bearing walls were surrounded by rows of elm tree which were very close to the external walls (1–1·50 m). The lighter elements, such as brick closures, verandas and pavements suffered sagging with settlements up to 5 cm, which was clearly associated with trees. The ground includes a bentonitic layer with $w_L = 150$–200% and swelling pressures up to 0·7 MPa.

Fig. 20. Pavement and basement disrupted by root penetration

Fig. 21. Subsidence cracking by desiccation

Fig. 22. Distorted pavement near adult trees

Fig. 23. Undamaged houses near a row of poplars (water supplied by a nearby sewage system)

Experience in other areas near Madrid have shown, however, that trees must be associated with soils of high swelling potential and buildings of very shallow foundation to cause significant damage. Fig. 23 shows a row of poplar trees very close to a building without any apparent effects even though the soil has $w_L = 80\%$ and $PI > 40\%$. The spread footing foundations were 0·80 m deep.

J. A. JIMÉNEZ-SALAS (Polytechnical University of Madrid)

Swelling clays are very common in Spain, and the capacity of vegetation to contribute to the production of damage on buildings is well researched. There is one aspect of its action that has not been described in the symposium.

One of the areas in Spain where this phenomenon is of widespread occurrence is Andalusia, in the south. Clays from the Upper Cretacic to the Pliocene, illitic and montmorillonitic, were deposited in a marine synclinal, and are now exposed in a heavily overconsolidated stiff state. They include a moderate content of calcium carbonate, very erratic, but with a high incidence into the 15–30% range.

Physical disintegration and leaching of the carbonates in the upper layers of the mass have caused expansive soils.

On the slopes, the thickness of the carbonate is limited by the erosion. In the valleys, the product of this erosion accumulates leading to vertisols of greater depth (local name 'buhedo').

However, in the upper layers of these vertisols, poor drainage, inverse leaching during the long dry season and conditions of bacteria life have led to the production of calcium carbonate enrichment, with concretions, and even in some places, a moderately cemented crust. These calcium carbonate enriched levels are, in a certain sense, stabilized and they do not have expansive properties.

The active zone in Andalusia is, generally speaking, not very deep—in practice, very seldom more than 3·5 m and it is infrequent for upheavals to occur in buildings provided with foundations more than 2·0 m deep.

This limited depth of the expansive layer produces a peculiar pattern of damage in the buildings (Jiménez-Salas, 1969). There is a predominance of horizontal effects, and cracks are very often strictly vertical.

But where trees have been planted, the roots extend in depth towards the permanently humid layers. These moist layers are located under the carbonated levels. These latter ones are mechanically disintegrated by the roots, and the calcium carbonate is changed into bicarbonate which is then absorbed by the root system. On excavation of these soils, as in road cuts for instance, it is very clearly seen that where a tree existed there is a deep dark 'pocket' penetrating across the whitish layers of carbonated clays. This increased thickness of the active layer is the main origin of the differential movements of the buildings in these cases.

Another cause of damage originated by trees is the heaving due to the roots extending below the foundation, looking for humidity, especially from leaking pipes and services.

My last point relates to the use of the ratio moisture content/liquid limit as an index of the state of desiccation of the soil. Driscoll writes that 'LL is preferred to PI or Plastic Limit (PL) since PL (and hence PI) values depend on a less reliable test procedure than that for LL'. This idea of the greater reliability of LL is put forward but the writer believes that it is only based on the primitive aspect of the PL determination technique.

For instance Lee, White & Ingles (1983) show that the coefficient of variation (CV) in the determination of LL oscillates, according to different authors, between 2% and 48%. However, the PL range is 9–29. The said authors recommend a standard mean value of the CV as 10% for both indices. It should be pointed out that these percentage values represent a much smaller absolute variation of the PL than in the LL, and as a consequence, LL has the greatest contribution to the uncertainties of the PI value.

But a more fundamental consideration is that NMC/LL represents an index that can be empirically correlated with the state of desiccation of a certain family of soils, in a similar position in relation to the 'A' line of Casagrande. But the ratio NMC/PL represents a real definition of the state of the soil, as at least in the Spanish climates (Thornthwaite index 0–40), NMC/PL oscillates between 1·0 and 0·6. The value 1·1 implies that the soil is only very slightly desiccated, and not able to swell dangerously. A value of NMC/PL of 0·8 predicts trouble. A much more detailed treatment of this point can be found in Jiménez-Salas (1980). In this reference there is also data about two other indices of desiccation. The first one is

$$\mathrm{DI(J\text{-}S)} = \frac{\log(\mathrm{PL/NMC})}{\log(\mathrm{PL/HM})}$$

where HM is the hygroscopic moisture at $\mathrm{pF}=6$, an unjustly neglected value, easily determined using a vacuum desiccator.

The other desiccation index, based on Livneh & Shlarsky (1965) data, is

$$\mathrm{DI(Liv)} = \frac{1{\cdot}9\mathrm{LL} - \mathrm{NMC} - 13{\cdot}47}{0{\cdot}28\mathrm{PL} + 2{\cdot}47}$$

The expansive clays in the surrounding area of Madrid are an example of how the NMC/LL ratio can be misleading. Very different from the southern Andalusian clays, they have been deposited into evaporitic lagoons, during the Miopliocene. They contain smectites and sepiolites in very variable proportions, and they can be on or well below the 'A' line (Jiménez-Salas, 1980).

In successive layers, it is possible to find clays with the same PL, say 30 for instance, and LL ranging from 85 to 45. If both layers have a moisture content of 24% their NMC/PL ratio will be 0·8 in both cases, their desiccation index $\mathrm{DI(Liv)} = 1{\cdot}75$ and their pF can be predicted with reasonable accuracy by the formula

$$\mathrm{pF} = 2{\cdot}2 + \mathrm{DI(Liv)} = 3{\cdot}95$$

In any case, the state of the layers in relation to consistency and state of desiccation, would be very similar. Nevertheless, their NMC/LL ratios would be 0·28 and 0·53, respectively. At least in this geological formation, the ratio NMC/LL is not representative of the real conditions.

T. I. LONGWORTH (Building Research Establishment)

The Building Research Establishment has recently collaborated with the Building Research Center of the Royal Scientific Society, Jordan, in a three-year study of building foundations on expansive clays in northern Jordan. The study was carried out in areas of Amman and Irbid where the presence of 2–10 m of superficial expansive clay has caused widespread building damage. The clay is very similar to the grumusolic soil of Israel described by Ravina. At the start of our study we anticipated that, due to the high temperature and aridity of the long Jordanian summers, the dominant ground movement affecting buildings would be heave arising from moisture diffusion following the covering of desiccated clay, as in South Africa (Williams & Pidgeon). However, our research has concluded that most cases of damage have resulted from ground shrinkage caused by moisture extraction by trees. An initial survey of more than 100 buildings damaged by ground movement in Amman and Irbid, found over 90% of them to have trees growing nearby. Typically, the pattern of cracks in walls and distortion of the floors has indicated a relative downward movement of the parts of the buildings nearest to the trees. Often trees had been planted in gardens around the buildings soon after construction, but damage did not occur until 10–15 years later, when the trees reached a height approximately equal to their distance from the building. Several other factors were found to have contributed in a lesser way to the occurrence and severity of the damage, including thickness of clay, type and age of construction, type and depth of foundations and excessive water infiltration. A discussion of these other aspects is given elsewhere by Longworth, Salem & Sunna (1984).

The principal tasks in our investigation of the problem were engineering geology mapping of the expansive clays, determination of the geotechnical properties in laboratory tests, and monitoring of ground movement, and building movement and damage in relation to climate and environmental changes at selected sites. The geological profile and geotechnical properties will be summarized briefly together with relevant comments; full details are given by Longworth *et al.* (1984) and Driscoll, Katkhuda, Salem & Longworth (1984). Also given is a typical case history of a damaged house monitored in Amman, which illustrates the complex effect of tree growth and tree removal on seasonal and long-term ground and building movements.

Geological profile and geotechnical properties

Examination of the superficial expansive clay in Amman and Irbid has shown it to be fairly uniform laterally, but to change in character with depth. A typical profile, given in Fig. 24, shows the principal lithological and structural

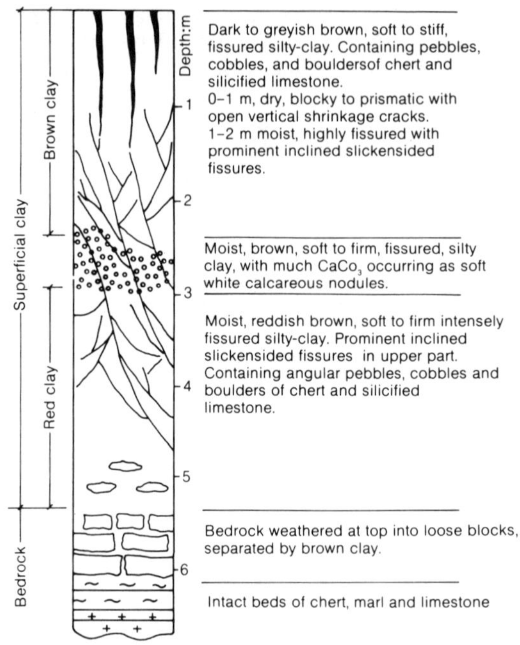

Fig. 24. Typical summer profile of superficial expansive clay

features. A colour change from greyish-brown to reddish-brown occurs at depths of 2–4 m, and corresponds to an increase in calcium carbonate ($CaCO_3$) content from less than 10% to over 20%. Additionally, there is usually a very obvi-

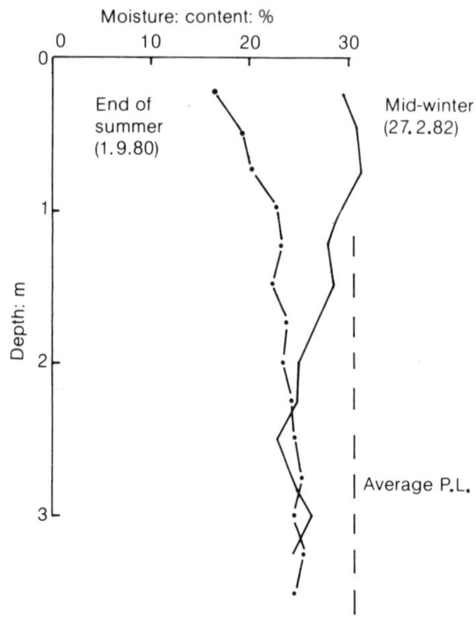

Fig. 25. Winter and summer moisture content profiles under grass-covered site in Amman

ous layer rich in $CaCO_3$ nodules at about 2·5 m depth.

The relative paucity of $CaCO_3$ in the near surface clay and its abundance in the nodule layer is considered to have resulted from weathering of the clay in a climate having pronounced wet and dry seasons. In the wet winters (rainfall 350–700 mm) of the northern highlands, rainwater dissolves $CaCO_3$ in the near surface clay and carries it down the soil profile. In summer much of the water is extracted directly from the ground by plant roots, leaving the $CaCO_3$ as a precipitate. Over a long period of time the $CaCO_3$ becomes concentrated as nodules at the base of the zone of annual moisture depletion and recharge. Visual examination of the soil profile thus indicates that typically the zone of annual moisture variation (the active zone) is about 2·5 m deep.

This is confirmed by moisture content profiles taken under an open grass-covered site in Amman, Fig. 25. The summer and winter profiles shown are not the extreme seasonal values, but they clearly indicate moisture variation to about 2·5 m depth. Similar occurrences of abundant $CaCO_3$ at the base of the active zone are evident in Fig. 9 of Richards Peter & Emerson, and Fig. 5 of Williams & Pidgeon. Recognition of this widespread correlation enables visual logging of the ground profile to give valuable information about the ground moisture regime.

The near-surface shrinkage cracks and deeper slickensided fissures shown in Fig. 25 are very similar to those described by Ravina. The fissures are in fact shear surfaces that have been produced in the laterally-confined deeper clay as the horizontal component of swelling is con-

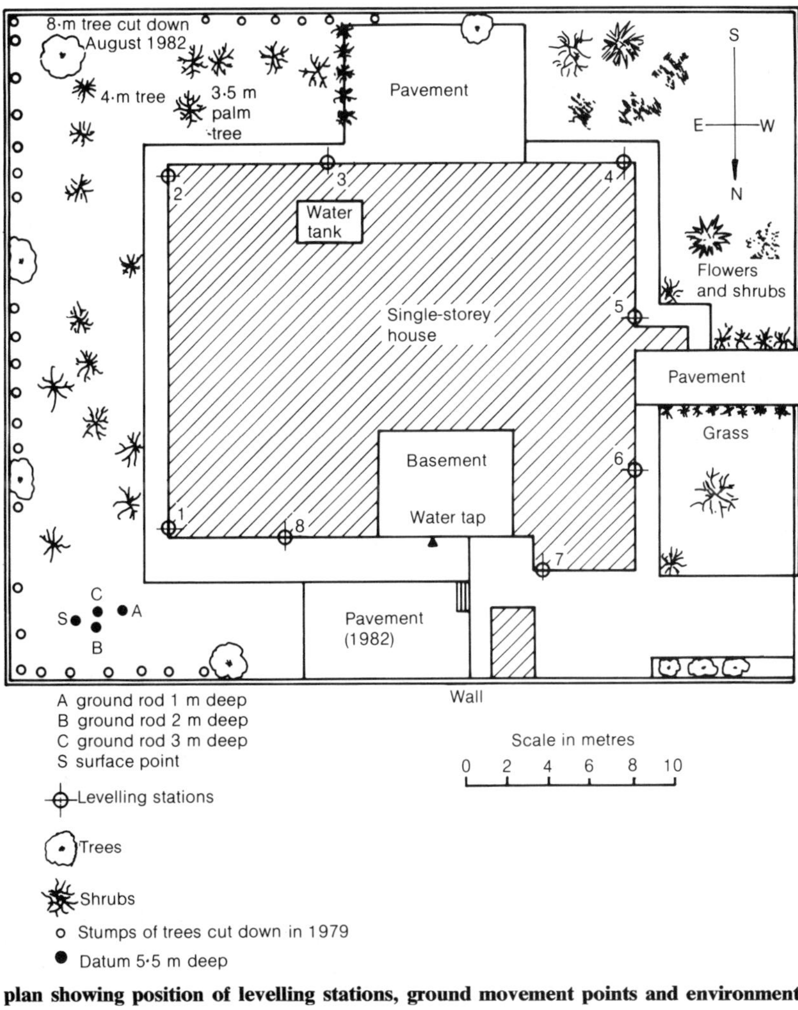

Fig. 26. Site plan showing position of levelling stations, ground movement points and environmental features

verted into vertical heave by passive failure. Blight & Williams (1971) have explained the mechanism analytically in terms of the changing effective stresses in the ground.

The geotechnical properties of the Amman superficial clay have been found to be nearly uniform with depth. Typical index properties of clay are: shrinkage limit 8–15%, plastic limit 28–35%, liquid limit 60–70% and clay fraction 55–60%. Plotting of plasticity index (LL-PL) against clay fraction on the Van der Merwe (1975) modified chart of expansiveness indicates the clay to be of high potential expansiveness. Constant volume swell tests on undisturbed samples have shown that swelling pressures in excess of 80–100 kPa can be developed in clay at 1–2 m depth when rewetted from initial moisture contents corresponding to the end-of-summer values shown in Fig. 25.

Tests on the Irbid clay, described by Driscoll *et al.* (1984), show fairly similar results, though with somewhat higher potential expansiveness.

Case history of house in Shmeisani, Amman

Figure 26 shows a plan of the single-storey house and surrounding garden. The house was constructed in 1970, with a reinforced concrete frame and concrete panels, partly clad in limestone blocks. The foundations of the inside walls consist of pads carrying columns, and those for the exterior walls are reinforced strips, 0·9–1·2 m wide, as in Fig. 27. Foundation depth is 2·5 m, while the depth of expansive clay on the site is about 6 m. The permanent water table is deeper than 50 m. Foundation pressure on the clay is about 50 kPa. Comparison with the geotechnical properties given above shows that this is less than the swelling pressures likely to be developed at foundation depths. The building is thus liable to heave movements resulting from wetting-up of the ground as well as settlement caused by drying out.

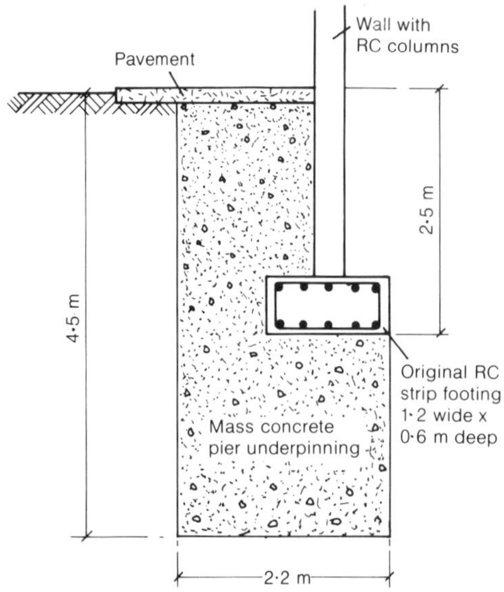

Fig. 27. Detail of original strip footing and 1979 underpinning

Some years after construction cracks resulting from ground movement appeared in the inside walls of the building. This damage became progressively worse and by 1979 cracking of both inside and outside walls was moderately severe (category 3; Building Research Establishment, 1981) with some cracks at the south-east corner open up to 20 mm. Remedial work was carried out in August 1979; this consisted of partial underpinning of the perimeter strip foundation to a depth of 4·5 m below ground level with mass concrete piers, as shown in Fig. 27. About 50% of the length of the east wall was underpinned, including the south-east and north-east corners. Since 1979 the house has continued to move, despite the underpinning; old cracks have reopened and new cracks have appeared.

Building and ground movements were measured over a 30 month period by precise levelling using techniques developed at BRE by Cheney (1973). Measurements were taken at monthly intervals relative to a stable datum founded on bedrock. Fig. 28 shows the movement observations in relation to the monthly rainfall of Amman.

Vertical ground movements measured at surface, 1 m, 2 m and 3 m depths are given in Fig. 28(c); and changes in thickness of the interposing ground layers (obtained by differencing the vertical movements) in Fig. 28(d). The ground layers down to 3 m depth show a seasonally cyclic pattern, with summer shrinkage and winter swelling. The seasonal ground movements, expressed as a percentage of the layer thickness, decrease from about 3% in the near-surface layer, to 0·5% in the 2–3 m depth layer. Below 3 m depth the gound shows a progressive swelling, totalling 1% or 25 mm over the period of observation. This swelling of the deeper clay can be explained by reference to the precise position and history of the ground in the area of the ground measurement points.

Figure 27 shows that this location at the north-east corner of the garden was originally surrounded by closely spaced 8 m high mature pine trees planted at the time of construction. The trees were cut down at the time of underpinning in autumn 1979. These trees, while growing, must have extracted large amounts of

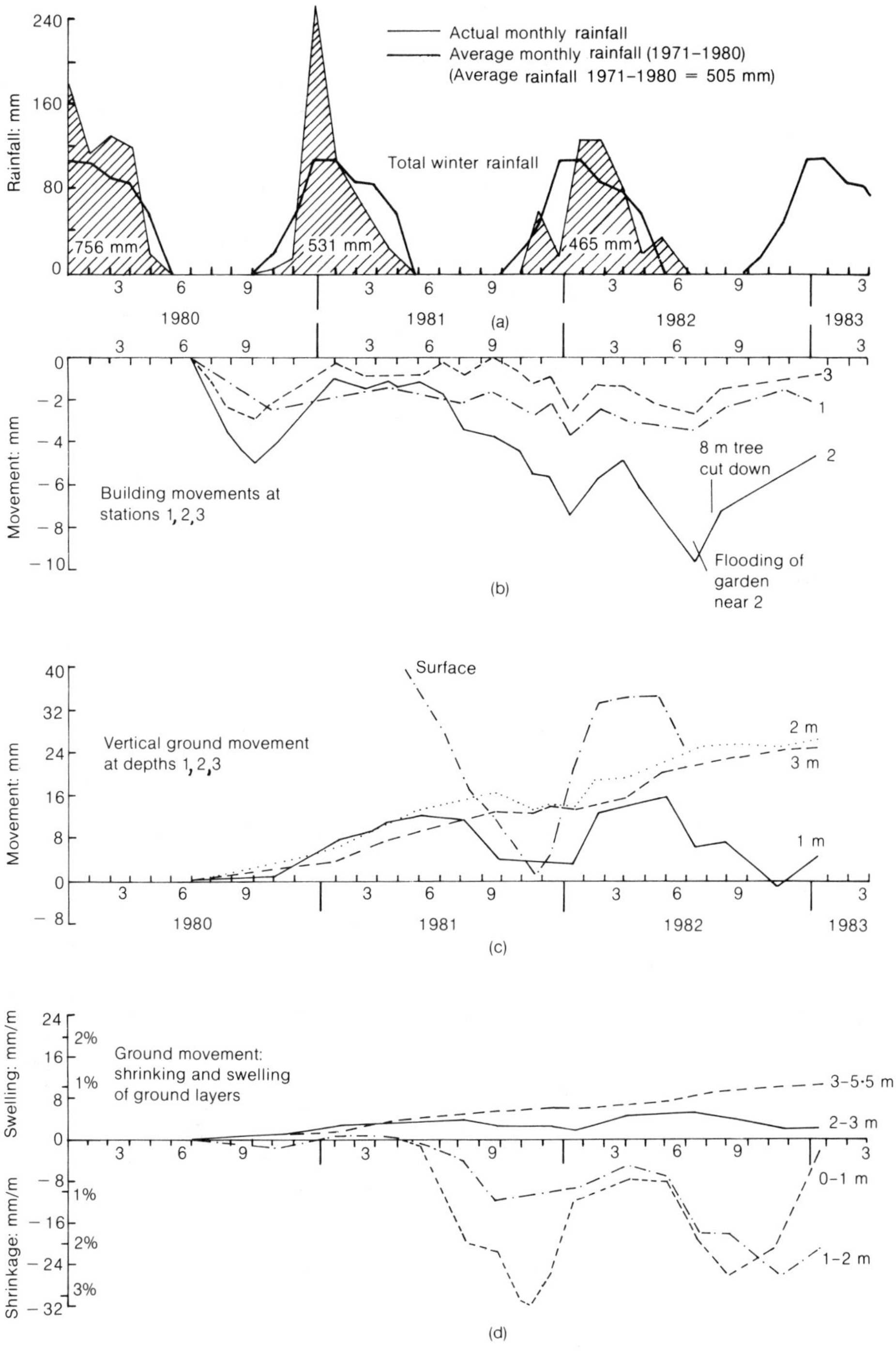

Fig. 28. Building and ground movements at a site in Amman

water from the clay in the deep seated rooting zone, causing shrinkage over many years. (These shrinkage movements probably caused building settlement, resulting in the damage observed in 1979). Removal of the pine trees has resulted in a 'rebound' swelling of the deeper clay layers as the moisture content has risen back to normal values. This recovery process has been aided by heavy watering of this section of the garden in summer.

Vertical building movements at levelling stations 1, 2 and 3 are given in Fig. 28(b); these stations show the maximum differential movement of the house. In general the measured movements of the building have been small and seasonally cyclic, moving up in winter and down in sumner, though the pattern has been apparently modified by frequency and intensity of summer garden watering in various parts of the garden. Typical movements, as at stations 1 and 3, are less than 4 mm. No significant new building damage was observed to be caused by these small seasonal movements around most of the building. However, new damage was observed at the underpinned south-east corner. Station 2 on this corner shows relatively large movements with settlements increasing to a maximum of 10 mm in mid-summer 1982. The resultant differential settlement was observed to cause previously repaired cracks (Fig. 29) to widen substantially in summer 1981 and again in summer 1982; the pattern of cracking confirming corner settlement. It is probable that this progressive settlement was caused by shrinkage of the clay around the south-east corner, resulting from the extraction of moisture by the roots of two adjacent trees (see Fig. 26). The proportionally greater settlement of 1982 is attributable to the less than average rainfall of winter 1981–82. Removal of the largest tree, an 8 m high eucalyptus, combined with a deliberate flooding of the surrounding garden in July and August 1982, led to an immediate recovery rise of several millimetres at station 2, though seasonally further settlement would have been expected.

Conclusions

The house described is typical of damaged buildings monitored in our study in that the strip foundation depths of 2·5 m have been sufficient to avoid significant damage from shallow seasonal ground movements caused by minor garden vegetation and modest garden watering. Both the original foundations and the partial underpinning to 4·5 m depth have, however, been insufficient to prevent substantial damage occurring in the longer-term due to tree activity. During the 30 months monitoring period, we observed evidence of both progressive settlement caused by tree growth, and progressive swelling caused by tree removal. These ground movements would appear to extend down to depths of nearly 5 m.

There are two possible reasons for the underpinned south-east corner to have settled during the period of observation. The first is clay

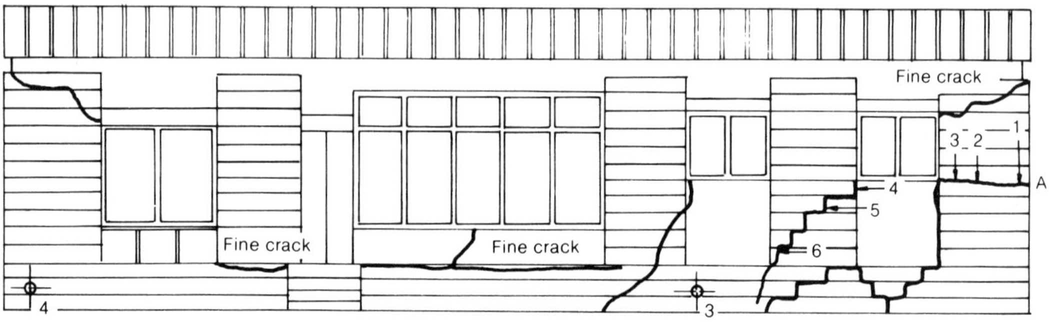

Crack width: mm				
Points	22.3.81	17.11.81	16.1.82	17.11.82
1	Fine	3.2	3.2	3.6
2	cracks	1.8	1.8	2.0
3	<0.5	<0.5	<0.5	0.5
4	<0.5	2.6	3.2	–
5	<0.5	1.5	1.6	–
6	<0.5	1.2	1.2	–

Fig. 29. South elevation of building showing crack damage. Typical observations indicate widening of crack A during summers 1981 and 1982

shrinkage due to moisture extraction by tree roots below the 4·5 m deep underpinning. The second is that underpinning and removal of many garden trees was carried out at the end of summer 1979 when the ground was seasonally desiccated. The following winter of 1979–80 was exceptionally wet and it is possible that relatively large swelling movements occurred, which lifted the building off the new mass concrete piers, leaving a void into which the uplifted foundations settled in the following drier years.

R. WYLES

In reply to Penman, if a local authority employs an independent consulting engineer to check on the designs of buildings submitted for Building Regulations approval, it follows that if proceedings are started against the local authority, they can bring in, as a third party to those proceedings, the engineer whom they employed. It is possible, of course, for the plaintiff to bring the engineer into the proceedings himself, but this would be unlikely to happen, as the plaintiff would usually be unaware that the local authority had employed a third party to check the design.

Under the Building Control Bill, which should receive Royal Assent in April 1984, there is a procedure whereby a developer can submit with his plans for Building Regulations approval, a certificate from an approved engineer or architect or surveyor certifying that the plans comply with the Building Regulations. If this procedure is adopted, the local authority, under the Bill, cannot query the design. It also follows that the local authority has no responsibility for approving the documents, and that in the event of failure later, the certifying engineer or architect would be responsible in place of the local authority.

With regard to Skilton's contribution, I consider that the part of the Acrecrest[3] judgement that dealt with the Dutch Elm disease is somewhat irrelevant, as it was decided by the Court of Appeal that, although the architects may not have considered the problems of heave in 1970 or 1971, they were well aware of the problems of subsidence and that being aware of those dangers, they should have considered the reverse situation, i.e. heave. I therefore agree that it would be unwise to rely, in any way, on the wording of this part of the judgement.

I think there may be a possible misconstruction of BS 3857, the code of practice for trees in relation to construction (British Standards Institution, 1980). The reference to '*where shrinkable clays are present and roots are found, or anticipated in the future*' relates to the duty of the designer to consider whether the roots of *existing* nearby trees might eventually find their way to the foundations of the property. The code does not, in my view (and indeed other professionals with whom I have discussed this problem), imply or infer that the designer should take into account the possibility that future owners may later plant trees.

The Bluett[4] judgement referred to slight or very slight cracks. What the judgement meant was that the judge accepted, because of the BRE classification and the fact that the damage had not been monitored, that substantial underpinning works were not necessary. The National House-Building Council and other professionals agree with the BRE scale of damage that very often unnecessary repairs have been carried out.

It does not seem reasonable that substantial repairs should be undertaken in order to remove blight when such repairs are unnecessary. Blight is a problem created not by owners but by building society surveyors and others. I note the reference to the household insurers, and this can of course be a problem. It is relevant, however, that even the household insurers are adopting the BRE classification, and have, in a number of cases, rejected claims because the damage is slight. My advice to intending purchasers, in order to get over this problem, is to always ensure that they take out their household insurance policy with the same company that is insuring the property for the current owner.

As far as the Ketteman[1] judgement is concerned, it should be remembered that, as in Bluett,[4] the claim failed against the local authority because there was no risk to health and safety, i.e. the damage was not sufficient to affect the structural stability or habitability of the dwelling. A claim against an architect or builder is, however, different from a claim against a local authority and a claimant may be able to recover damages from an architect or contractor which would not be recoverable from a local authority. The damage in the Ketteman properties was worse than in the Bluett case, although I would agree that it was perhaps arguable that substantial repairs were necessary. The Ketteman judgement is in fact going to the Court of Appeal in June 1984.

To answer Statham & Thomas, the BSI code of practice (British Standards Institution, 1980) does not mean that engineers have to take into account the fact that trees may be planted later. What they do need to take into account are the possible effects on the foundations from nearby trees whose roots have not yet penetrated as far as the site where the foundations are to be laid.

With reference to Cox's question, Digest 251

(Building Research Establishment, 1981) does not pretend to be all-embracing, and soil type location, etc. must all be taken into account in determining what repairs are necessary.

With reference to Train's comments, the BSI code of practice (British Standards Institution, 1980) does not mean that engineers have to take into account the fact that trees may be planted later. What they do need to take into account are the possible effects on the foundations from nearby trees whose roots have not yet penetrated as far as the site in question. If an owner plants trees later, after the property has been completed, then he could not sue the engineer or contractor, although doubtless he would have a claim under his household insurance policy for subsidence or heave (if the converse happened and he cut down an existing tree).

It would not be feasible to append Practice Note 3 (National House-Building Council, 1969) to the deeds of a property, although I think that Train's suggestion about NHBC publicising this problem has merit. I have referred it to our Technical Department to see whether any of the information booklets prepared by NHBC could take this point into account.

To answer Leach, where trees have been planted along road verges by the local authority, the local authority is responsible for any tree root damage (in nuisance). Local authorities pay out substantial amounts each year in respect of tree root claims. I believe this to be the case even where the house has been built later than the tree was planted. Even so, I would agree that if the builder does not take into account Practice Note 3 (National House-Building Council, 1969) in constructing a dwelling, he would be liable to the owner if damage occurs later.

R. DRISCOLL

Sharrock advocates the use of the shrinkage limit (SL) to detect desiccation in clays. However, apart from indicating approximately the moisture content below which shrinkage of the soil is insignificant, comparing 'natural' moisture content (NMC) of a clay with SL is no better a means of detecting desiccation than comparison with other index properties. As Sharrock implies, there are various laboratory techniques for measuring porewater suction in soil samples. However, none is known to the Author to be readily available outside research laboratories.

Wilson advocates the use of consistency index, I_c, to indicate the state of desiccation. He points out that the index has values greater than 1 when NMC is less than PL.

However, as an indication of desiccation it is no more reliable than is 0·4 LL since at pF = 2, values of I_c from Table 7 range between 0·7 and 0·88, while at pF = 3 values range between 0·8 and 0·96. Wilson rightly concludes that no one test is sufficient to confirm desiccation or to differentiate between desiccation and a low NMC due to overconsolidation. He also concludes that comparison of two sets of values, one within the zone expected to be desiccated and the other where desiccation is not envisaged, is the only conclusive method of assessment short of more sophisticated soil tests. These two points are also made by Statham & Thomas, and Reeve in their discussion contributions.

Statham & Thomas rightly state the assumptions of $A = 1/3$ and $B = 1$ in the calculation of suction generated upon sampling. In the example presented in the paper a saturated sample is implicit. The example was intended to demonstrate that, despite sampling below a water table, swelling could still be measured in the oedometer.

The Author agrees that the use of 0·4 LL to indicate significant desiccation does not distinguish between desiccation and overconsolidation. It would be of benefit to incorporate an overconsolidation ratio into the indicator but this is not available for the data presented in the Paper. The Author agrees that no one soil parameter may be taken in isolation to give a definite indication of desiccation.

McEntee presents interesting, if brief, case histories which raise many questions including: how and why did the deep strip foundations fail? why did the piled foundations fail? how deep is the granular blanket beneath the raft foundation? It seems unlikely that the reinforced concrete slab with down-stand edge beams described could prevent cracking from the 60 mm differential movement indicated in Fig. 12.

In his discussion contribution Reeve raises some valid points about predicting shrinkage potential from soil properties. He confirms findings made by many researchers that shrinkage correlates relatively well with LL or PI. While shrinkage is seen to have a relatively high negative correlation with dry density (γ_d), γ_d has been found elsewhere to be positively correlated with swelling.

Reeve's comment on the need to consider volume change over a specific soil suction range is very relevant and reinforces the point that assessment of volume change potential from index and other soil properties serves to provide only a broad indication of field events.

In his discussion, Leach asks if appropriate water shrinkage factors $\Delta w/\Delta z$ can be allocated to the four soil categories in Table 2 of the

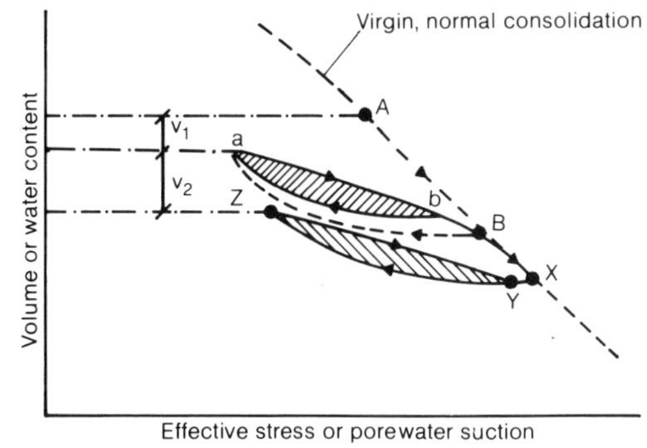

Fig. 30. Change in soil volume with increasing effective stress or pore water suction

Paper. The Paper states that $\Delta w/\Delta z$ values of both 4 and 8 have been measured for 'high' to 'very high' shrinkage potential soils, indicating the lack of precision inherent in predicting ground movements. Clearly, a great deal more research of the type described by Biddle in his paper, with the additional measurements of ground movements, is required to improve matters.

A number of cases of substantial structural crack recovery in the winter following dry summer opening have been encountered by the Author, but recovery has never been found to be complete. The recovery has two elements: the soil recovery and the structure recovery. In the former, the degree of recovery depends crucially on the stress history of the soil (as well as on the swell potential embodied in the clay mineralogy and expressed in index properties).

This point can best be illustrated in Fig. 30 which shows the volume change of a soil with changing effective stress, or porewater suction. If the clay was originally normally consolidated to a point A, say, and then experienced its first dry summer, the increasing porewater suction, or effective stress, would reduce its volume to, say, point B.

In the following winter the soil would swell, say to point a, with a consequent permanent reduction v_1 in soil volume. Subsequent summers might dry the soil to point b, with winter swelling back to a, with virtually complete recovery of volume. When a summer drier than hitherto experienced occurred, the soil volume would shrink to, say, X, recovering not to point a but to Z the following winter, with a permanent volume decrease, v_2. Subsequent seasons would closely follow the hysteresis loop YZ.

The structural element of recovery depends largely on how well the two fracture surfaces are able to fit together. With regard to heave of shallow foundations following tree removal, it is the very lack of vertical confining pressure that enables the soil to swell preferentially more in the vertical direction.

Rodriguez Ortiz & Serrano present some interesting information from Spain, in particular that damage is rare when foundations exceed 1 m in depth and that their risk index is zero when the distance between tree and building exceeds 10 m.

In another contribution from Spain, Jiménez-Salas discusses various desiccation indices based mainly on Atterburg Limits. He concludes that for Spanish conditions a ratio of NMC/PL of 0·8 or less indicates desiccation of sufficient magnitude to generate swelling problems.

Cox raises doubts about a recommended damage classification (Building Research Establishment, 1981). The BRE Digest 251 goes to some lengths to explain how the locations and types of cracking in low-rise buildings should be sketched and described. Additionally, the measurement of distortions, tilts and slopes is recommended as well as a 'thorough description' of the materials used in construction. Though the Digest is aimed primarily at damage in masonry construction on clay soils, since these constitute the majority of cases, the many possible causes of cracking of building fabrics and foundations movement are described.

P. G. BIDDLE

Many of the comments note that the majority of results presented in the Paper refer to seasonal changes in moisture content, and question their applicability if permanent deficits also exist. This aspect was covered in the discussion on the influence of soil permeability and permanent water deficits, which records that permanent deficits were only observed with the nine poplar trees. These permanent deficits were assessed by assuming that the springtime readings at the comparison hole provide a measure of normal field capacity, and contrasting this profile with those close to the tree. Such measurements are directly similar to the single set of results presented by Cox for the ash tree in Regents Park (see Fig. 5).

Although the Paper asserts that no semi-permanent deficits could be demonstrated for any of the other species apart from poplar, further analysis of the 1983 results has indicated that slight deficits might persist near some lime and horse chestnut trees. However, the extent of these is not as great as the permanent deficits near poplar.

Some of the comments refer to previous research or comment on the extent of deficits for unspecified species of tree. This omission of information about the species is regrettable when the Paper places so much emphasis on its significance. In this respect it is interesting to note that the two case histories of heave quoted by McEntee refer to elm and oak trees, which are two other species with a high water demand.

The two case histories of heave illustrate this problem which is becoming increasingly recognized as a cause of structural failure. There are two aspects of these cases that require further comment. Firstly, the heave of the terrace of four houses is attributed to the culverting of a stream which deprived the oak trees of water and the subsequent ingress of flood water into the foundations. Such causes do not explain the very rapid development of damage within 6 months of the houses being occupied. An alternative explanation, and one which is frequently observed in other situations, is that the excavation for the deep strip footings severed the roots which passed beneath the building. This will have allowed the soil trapped within the box-like foundations to rehydrate and swell, causing the 10 mm horizontal movement as well as the uplift.

Secondly, Fig. 12 shows the monitoring of swelling (heave) of two levelling stations on a house near elm trees that were felled. The time axis of this graph uses a log scale which appears to indicate no effect after removal of the first two trees but a very rapid response after the final tree was removed. However, if this graph is plotted on a normal time scale, it can be appreciated more easily that the final tree was removed after 2 years. Rapid heave continued for four years and then at a gradually diminishing rate for about another 10 years.

Wilson and others comment on the scope of the investigation, and suggest that it should include other aspects (i.e. additional species such as oak, measurements of soil movement, descriptions of soil profiles, correlation with normal moisture content data, and situations such as competition with other trees and the influence of covered areas). All of these aspects are being covered in the existing projects, but lack of space prevented their inclusion. Furthermore, the investigations into 36 trees described in the Paper have now been supplemented by an additional 24 trees under funding from the Department of the Environment. This extension of the project includes oak, Norway maple and whitebeam as additional species, and experiments on the effects of pruning on the water demand of plane trees. It is also suggested that the measurement of moisture deficit should be correlated with direct measurement of the soil suctions as described in the Paper to the Symposium by Richards *et al.* The results of their research illustrate the many problems of this approach, particularly with the lower suction pressures produced by most species in this country. However, it is appreciated that such correlations could be useful, and consideration will be given to including such work in the future.

The study of the existing 60 trees will continue for at least the next two summers, and future publications will describe the results in greater detail than was possible in this short Paper. This mass of data for 60 trees over many seasons must provide a far more sound basis for future guidelines, compared to the existing guidelines which are based on a minimum of factual evidence. It is to be hoped that one of these summers will produce a reasonable drought but this is not essential. The seasonal development of the moisture deficit is being correlated with the weather conditions; even a drought of moderate severity such as in 1983 should permit extrapolation to more extreme conditions.

If nothing else, this data should provide a factual basis for checking the mathematical flow modelling systems which are suggested by Cox. Ultimately his proposals may be feasible but the problem is far more complex than he indicates. Uptake by trees is not analogous to drawdown to a point source such as a well or sump; the tree root system is a vast, dynamic, sponge-like organism which responds to many stimuli apart from moisture availability. At present it is difficult to envisage any model capable of predicting the behaviour of a system that is still so poorly understood.

B. G. RICHARDS, P. PETER and W. W. EMERSON

The discussion by Cox has mentioned the technique of using flow nets (or similar mathematical models) to establish the flow from the water table to a tree (or line of trees). It is true that mathematical models are providing a useful insight into the physical mechanisms operating. However, traditional flow nets have been found to be very inaccurate in highly expansive clays in which soil suctions exist. This is due to the highly non-linear nature of the flow properties (e.g. permeability), hysteretic behaviour and structural effects (e.g. cracks, fissures). The Authors are in fact using mathematical models extensively, but these are non-linear transient finite element models, which attempt to consider the factors mentioned. These models are very complex and cannot be considered at this stage as practical techniques.

In reply to Sharrock, the shrinkage limit test was considered many years ago, but in comparison to the procedures described in the Paper, it was found to be very variable in expansive clays in Australia and gave poor correlations with performance. We, for example, exclusively use soil suction for monitoring soil moisture conditions. The question therefore arises as to the relevance of shrinkage limit values when using soil suction. As an example, the soil suction at the shrinkage limit of the red-brown earth subsoil at 0·70 m (Fig. 9 of the Paper) was reported by Holmes (1955) to be pF 5·7, whereas the value for a black earth at 0·5 m depth (similar profile to Fig. 7a of the Paper) was greater than pF 6·0. The value of the maximum pF to which a subsoil can be reduced by vegetation is taken to be 4·2 (the conventional wilting point), although it will be noted that immediately under gum trees, the pF can be increased to 4·5 or 3·3 MPa (Fig. 10 of the Paper).

The Authors would agree with Sharrock that soil suction is 'of more importance'. One major difference between Australian and English conditions is the frequent presence of high salt concentrations in the pore water. For example 85% of the total in situ suction measured on the black earth subsoil is due to solutes (Pile, 1980). In such cases, an estimate of matrix suction only, as obtained from filter papers in liquid contact with the soil would be inadequate information. In spite of this, the filter paper method is widely used in Australia as a cheap, simple approximate method by engineering consultants. Because of confusion as to what components of soil suction it actually measures, the filter paper method cannot be considered an accurate method. The glycol osmotic method has been developed commercially in Australia. But due to its cost and unsatisfactory performance in both the laboratory and the field, its manufacture has been discontinued.

Measurement of total suction is particularly useful in non-saline soils, where its value is much less sensitive to changes in texture than water content (Fig. 2, Emerson & Peter, 1981). Also the change in total suction of unconfined cores for a given amount of linear shrinkage can be used in the laboratory to calculate approximate values of the instability index $I_{p''}$ (Peter, 1979). Then potential movement due to drying by vegetation or ponding by water can be easily calculated. Note that whether solute suction will contribute towards the swelling of a clay soil can be determined by a simple dispersion test, as indicated in the Paper.

The measurement of total suction is now carried out as a routine procedure by the Authors. Two psychrometric methods are used, one based on thermistors and the other on thermocouples. While the capital cost of psychrometric apparatus is high, the cost per test is very low when the apparatus is in continued routine use. Therefore, if one has access to such apparatus, its use is to be preferred to the other techniques mentioned.

With reference to the contribution by Rodriguez, Ortiz & Serrano, the Authors suggest that conifers cause less damage than deciduous trees. This is most unlikely to be due to a lower demand for water, but rather that perennial trees can make use of the artesian water at 9 m. In Fig. 15 of the Paper we report evidence to show that mature pine trees can utilize water at this depth. It would be interesting to know the salt content of the artesian water. The pF value of above 4 reported to occur in the soil close to the water table may be due in large part to salts. Some pines, such as *Pinus pinaster* are salt tolerant.

With regard to the contribution by Longworth, the suggestion that a layer of nodular

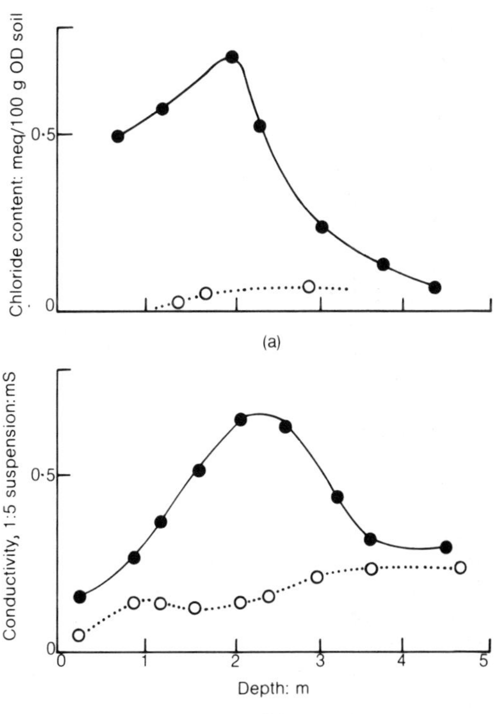

Fig. 31. (a) conductivity (b) chloride profiles in a red-brown earth (RB3) 2 m (o) and 70 m (o) away from a *Pinus palustris*, 21 m high

carbonate corresponds to the base of the zone of annual moisture depletion and recharge must be treated with caution. In the first instance the carbonate layer may be part of an earlier buried soil. In general, the depth of seasonal wetting in a semi-arid environment depends on the vegetation. Under pasture, a red-brown earth profile is completely wetted up by winter rains, whereas under gum trees the limit of water penetration is about 2·4 m (see Fig. 9 of the Paper).

In such profiles where average salinity is low, the average depth of wetting is easily picked up by the change in electrical conductivity or chloride content with depth. Fig. 31 shows the contrasting profiles under pasture and under *Pinus palustris* (see Fig. 10 of the Paper). The average depth of wetting is 2 m. The corresponding profile given in Emerson & Peter (1981) shows that the maximum carbonate occurs at 0·9 m.

A. A. B. WILLIAMS and J. T. PIDGEON

With regard to the contribution by Coatsworth & Evans, the writers seem to have misunderstood the Authors' use of the phrase 'easily visible roots' which in fact appears to be the same as their 'very fine roots'. In the study of a soil profile down a large diameter borehole a particular point is made of recording the existence of roots, or the presence of ants, or termite workings. Further, it is just as important to include a comment such as 'no sign of roots' to ensure the proper engineering interpretation of a soil profile description. Most of the records of site investigations referred to by the Authors correspond to sites with grass vegetation rather than trees and consequently root depths tend to be more limited. In the case-record concerning gum tree roots, however, the Authors findings seemed to be confirmed by the writers. The position of the water table may indeed be strongly influenced by the vegetation, and there are many records of groundwater levels being lowered by the presence of gum plantations, but this does not mean that the roots must extend to the water table.

The measurement of soil suction is becoming more common with the advent of commercially available psychrometers and several consulting firms and statutory organizations have now acquired the equipment necessary for laboratory or field work. This technique is viewed as providing a useful site investigation tool and not merely a method of predicting heave using semi-empirical relationships. It is perhaps unwise to generalize on the suitability or accuracy of the heave/suction relationships proposed by Brackley or Johnson & Desai. The former has been found to provide acceptably good results at a number of sites in South Africa but these results require further verification before universal application is advocated. The undesirability of using the plasticity index as the material parameter representing a measure of the physico-chemical properties of the soil is acknowledged, and it is felt that some parameter expressing the fabric of the soil should be incorporated in the equation for predicting swell. The use of swell under load tests in the laboratory shows promise if severe sampling disturbance can be overcome. Research is currently being undertaken on the development of a sampling tube that will enable the retrieval and testing of soil samples without disturbing the lateral stress conditions existing in situ.

Rodriguez Ortiz & Serrano and Jiménez-Salas describe typical damage to light structures due to 'shrinkable' or to 'heaving' clays and it was not realized that the problem was so widespread in Spain. While being aware of the lack of wisdom sometimes in over-simplification, continual rise or heaving type of movement in South Africa is usually associated with the desert, steppe or humid mesothermal climates all having the dry season in winter. The cyclic type of movement due to swelling and shrinking is usually encountered in the humid mesothermal climatic regions where the dry season is in summer, which gives a much greater seasonal variation of effective moisture. In fact this large seasonal variation is only found where the seasonal march of precipitation is opposite to that of the water need, or potential evapotranspiration. Similar climatic regions are encountered in Spain and it would be interesting to relate the actual pattern of building movement to the seasonal water balance at various Spanish stations. (Thornthwaite, 1948). An important observation was also made concerning the remarkably high values of soil suction encountered only a little above the water table at 9 m depth. This emphasizes the important point that the effective stress in the soil may bear relationship to any extrapolation of the hydrostatic positive pore pressure condition below a water table when dealing with a site subject to the recent effects of vegetation.

Longworth has mentioned that the cyclic pattern of foundation movement observed in northern Jordan was different from that expected. However from Fig. 14 it is noted that this region would also be classified as having a steppe climate but with the dry season in summer. Again, a study of the annual march of rainfall and precipitation in terms of Thornthwaite's concept would give a good idea of monthly variation in soil moisture.

The writer makes a most interesting comment about the presence of calcrete nodules near the base of the zone of annual moisture under natural conditions. However, he is reminded of the Authors' comments about the extent of the active soil zone in regions where desiccation may persist to much greater depths than that of seasonal variation. The measurement of soil suction as well as the moisture content profiles would aid in understanding the anomalies in movement of foundations that seem to have arisen even after underpinning.

I. RAVINA

In answer to Cox, indeed a blocky structure with slickensides of about 45° inclination is observed in the deeper layers of most expansive soil profiles. The observed sheaths of compressed oriented clay particles consisting of the surfaces of such fissures, as mentioned in the Paper, may be evidence for the sliding concept suggested by the writer. His approach is generally plausible supporting the observations of larger cracks forming at boundaries of zones of different moisture changes. It is noteworthy that cracking is strongly dependent also upon the rate of moisture change and not only its magnitude.

W. G. HOLTZ

The contribution from Rodriguez Ortiz & Serrano provides an excellent description of desiccation damage from montmorillonite-type soils. However, we still have the problem of educating the 'public' about the damage to buildings and other structures that can develop from the shrinking and swelling of critical soils. We have been trying to do this in several parts of the United States by means of publicly oriented symposia. For these symposia the 'public' is not limited to the owners or prospective owners of homes or buildings, but very importantly includes general civil engineers, architects, builders, city engineers, building inspectors, land developers, mortgage companies and real estate agents and companies.

The contributors mention that their damage index could be improved by including a rating of soil sensitivity to moisture–volume changes. The writer believes this would be a necessity. We have found that clay soils with liquid limits less than 40 and shrinkage limits greater than about 16 very seldomly cause significant damage from shrinking or swelling. (see Holtz & Gibbs 1956). The contributors' index should be compared with criteria developed at the Texas A & M University, College Station Texas, which would be of interest to us all.

We have found that spread footings, normally, are too shallow to prevent desiccation damage from large water-loving trees. To combat such problems deeper caisson footings are often used. The remedial measures recommended in the discussion are important, but should include a criteria of tree distance from buildings for the various types of trees.

With regard to Longworth's contribution, the damage to buildings and the nature of the soils described are good case histories of damage similar to that reported for so much of the world. Each case history is of value to the geotechnical engineering profession.

The discussion of the calcium carbonate ($CaCO_3$) nodules is of considerable interest. I have seen this occurrence in parts of the United States, Columbia, Israel, Peru and other countries. It almost seems like this occurrence is an anomaly. If the calcite, of rather low solubility, was broken down to its components and dispersed it would control the expansiveness of the clays. Expansive clays treated with lime, CaO or $Ca(OH)_2$, can be rendered almost harmless (Holtz, 1969).

It is interesting that the contributor has used the nodule concentration zone as an indicator of the zone of moisture depletion and recharge, as have some other investigators.

CASES

1. *Ketteman* v. *Hansel Properties Ltd* (1983). Unreported.
2. *Pirelli General Cable Works Limited* v. *Oscar Faber & Partners* (1983). 1 All England Law Reports.
3. *Acrecrest Ltd* v. *W. S. Hattrell & Partners* (1983). 1 All England Law Reports.
4. *Bluett and Another* v. *Woodspring District Council* (1982). Estate Times, May.

REFERENCES

Anderson, M. G. (1979). On the potential of radiography to aid studies of hillslope hydrology. *Earth Surface Processes* **4,** 77–83.

Anderson, M. G. & Kneale, P. E. (1980a). Porewater changes in a road embankment. *Highway Engineer* **25,** 11–17.

Anderson, M. G. & Kneale, P. E. (1980b). Porewater pressure and stability conditions on a motorway embankment. *Earth Surface Processes* **5,** 37–46.

Blight, G. E. & Williams, A. A. B. (1971). The formation of cracks and fissures by shrinkage and swelling. *Proc. 5th Reg. Conf. Africa Soil Mech., Luanda.*

Brackley, I. J. A. (1975). *The interrelationship of the factors affecting heave of an expansive, unsaturated, clay soil,* pp. 159–166. PhD thesis, University of London.

Brackley, I. J. A. (1980). Prediction of soil heave from suction measurements. *Proc. 7th Reg. Afr. Conf. Soil Mech., Accra* **1,** 159–166.

British Standards Institution (1980). *Code of practice for trees in relation to construction.* BS 3587. London: British Standards Institution.

Building Regulations (1976). London: HMSO.

Building Research Establishment (1980a). *Low rise buildings on shrinkable claysoils: part 1.* Digest 240. London: HMSO.

Building Research Establishment (1980b). *Low rise buildings on shrinkable claysoils: part 2.* Digest 241. London: HMSO.

Building Research Establishment (1981). *Assessment of damage in low-rise buildings.* Digest 251. London: HMSO.

Cheney, J. E. (1973). Techniques and equipment using the surveyor's level for accurate measurement of building movement. *Symp. Field Instrumentation.* London: British Geotechnical Society.

Cutler, D. F. & Richardson, I. B. K. (1981). *Tree roots and buildings,* p. 94. Harlow: Construction Press.

Driscoll, R. Katkhuda, I. E. D., Salem, A. & Longworth, T. I. (1984). Geotechnical properties of expansive clay soils in northern Jordan. *Proc. 5th Int. Conf. Expansive Soils, Adelaide.* (To be published).

Emerson, W. W. & Peter, P. (1981). The engineering properties of red-brown earths. *Prescott symposium on red-brown earths of Australia,* pp. 149–163. Adelaide: Commonwealth Scientific and Industrial Research Organization.

Holmes, J. W. (1955). Water sorption and swelling of clay blocks. *J. Soil Sci.* **6,** 200–208.

Holtz, W. G. & Gibbs, H. J. (1956). Engineering properties of expansive clays. *Trans. Am. Soc. Civ. Engrs* **121,** 641.

Holtz, W. G. (1969). Volume change in expansive clay soils and control by lime treatment. *Proc. 2nd Int. Conf. Expansive Clay Soils.*

Johnson, L. D. & Snethen, D. R. (1978). Prediction of potential heave of swelling soil. *Geotech. Testing J.* **1,** No. 3, 117–124.

Jiménez-Salas, J. A. (1969). Discussion on the Speciality Session on Expansive Clays. *Proc. Int. Conf. Soil Mech. Mexico.*

Jiménez-Salas, J. A. *et al.* (1980). *Geotecnia y cimientos 3.* Ch. 5, pp. 535–650, Madrid: Rueda.

Lee, I. K., White, W. & Ingles, O. G. (1983). *Geotechnical engineering.* London: Pitman.

Livneh, M. & Shlarsky, E. (1965). Saturation criteria in pavement design for semi-arid zones. *Moisture equilibria and moisture changes,* Australia: Butterworths.

Longworth, T. I., Salem, A. & Sunna, B. (1984). Correlation of building damage on expansive clay in Jordan with geology, local environment and type of construction. *Proc. 5th Int. Conf. Expansive Soils, Adelaide.* (To be published).

National House-Building Council (1969). *Root damage by trees—siting of dwellings and special precautions.* Practice Note 3.

Peter, P. (1979). Soil moisture suction. In: *Footings and foundations for small buildings in arid climates* (eds P. J. Fargher, J. A. Woodburn and J. Selby) pp. 46–62. Australia, S. A. Division: Institute of Engineers.

Pile, K. C. (1980). The relationship between matrix and solute suction, swelling pressure and magnitude of swelling of reactive clays. *Proc. 9th Aust.–N. A. Conf. Soil Mech., Auckland* **1,** 197–206.

Reeve, M. J. Hall, D. G. M. & Bullock, P. (1980). The effect of soil composition and environmental factors on the shrinkage of some clayey British soils. *J. Soil Sci.* **31,** 429–442.

Thornthwaite, C. W. (1948). An approach toward a rational classification of climate. *Geogr. Rev.* **38,** 55–94.

Ward, W. H. (1953). Soil movement and weather *Proc. 3rd Int. Conf. Soil Mech., Zurich* **4,** 477–482.

Van der Merwe, D. H. (1975). Contribution to speciality session B: current theory and practice for building on expansive clays. *Proc. 6th Reg. Conf. Africa Soil Mech., Durban* **2,** 166–167.

BIBLIOGRAPHY

Holtz, W. G. & Hart, S. S. (1978). Home construction on shrinking and swelling soils. *Proc. Conf. Home Construction on Shrinking and Swelling Soils, Denver.* Available from ASC, National Science Foundation, Washington or Colorado Geological Survey, Denver.

Jones, D. E., Jr & Holtz, W. G. (1973). *Expansive soils—the hidden disaster.* American Society of Civil Engineers.